Philip Hoare is the author of five books of non-fiction, most recently *England's Lost Eden: Adventures in a Victorian Utopia*. He lives in Southampton.

From the reviews of *Leviathan or, The Whale*:

'Insights and images rise in plumes from almost every page'
Jonathan Bate, *Sunday Telegraph*

'This history of man's dealings with whales is respectful, even mystical'
Peter Lewis, *Daily Mail*

'An enjoyable trawl through the history, literature and lore of whales . . . As well as being a showcase for descriptive prose of great beauty, *Leviathan* is full of fascinating facts'
Ian Pindar, *Guardian*

'So compelling and all-encompassing that it cast a spell on me that endured for days after I had done turning its beautifully illustrated pages . . . This is the book [Hoare] was born to write, a classic of its kind. What poetry there is here and what a balm for the soul'
Rachel Cooke, *Observer*

'A scintillating, scattershot, blunderbuss of a book. Throughout the book, Hoare's unbridled enthusiasm for his subject is infectious . . . this thoroughly engaging, rigorously researched and often revelatory book is a joy to read and one which Melville, surely, would have appreciated'
Doug Johnstone, *Independent on Sunday*

'Hoare's idiosyncratic mingling of autobiography, anthropology and archaeology has reached its zenith . . . an enthralling volume
Andy Miller, *Daily Telegraph*

'An elegant writer with a sharp eye for quirky detail . . . A lyrical and timely reminder of what we have to lose if we don't change our greedy ways'
Sara Wheeler, *Mail on Sunday*

'In Hoare's hands, whales are al
interesting'

'Richly stocked with whale lore and written with admirable intensity and élan . . . Shuttling between inhuman actuality and anthropomorphism, Hoare breaches the surface of his subject in the most profound fashion' Brian Dillon, *Irish Times*

'Hoare's wonderful illustrated biography of this most magnificent beast is studded with glittering shards of natural history and social science' Claire Allfree, *Metro London*

'The author's passion for whales is infectious'
Esquire magazine

'Hoare's personal pilgrimage, wandering, reflective, frequently very personal, owes much to W. G. Sebald, including the device of peppering the text with black and white pictures. Whales have a very intimate and troubled relationship to man, one which this elegiac book does much to illuminate'
Martin Latham, *Waterstone's Books Quarterly*

'A wonderfully idiosyncratic book, passionate zoology counterpointed with the glories of *Moby-Dick*. This is a deep book about the deep: an inspiring book about inspirational beings. We need the wild world more with every passing piece of destruction. If you can't board a ship this week, read this book instead'
Simon Barnes, *The Times*

'A compendious and visionary account of the author's cetacean romance, and a rigorous reading of Herman Melville's *Moby-Dick* . . . Hoare is a cultural historian of consummate skill'
Frieze

BY THE SAME AUTHOR

Serious Pleasures: The Life of Stephen Tennant

Noël Coward: A Biography

Wilde's Last Stand: Decadence, Conspiracy and the First World War

Spike Island: The Memory of a Military Hospital

England's Lost Eden: Adventures in a Victorian Utopia

LEVIATHAN

or,

The Whale

Philip Hoare

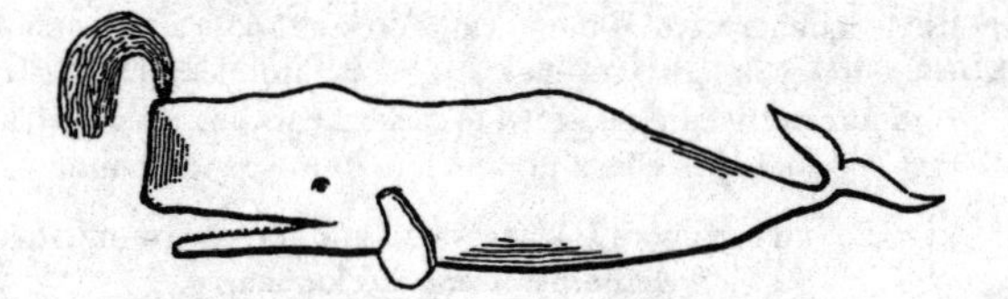

FOURTH ESTATE • *London*

Fourth Estate
An imprint of HarperCollins*Publishers*
77–85 Fulham Palace Road
London W6 8JB

www.4thestate.co.uk
Visit our authors' blog at www.fifthestate.co.uk
Love this book? www.bookarmy.com

This Fourth Estate paperback edition published 2009
3

First published in hardback by Fourth Estate in 2008

A catalogue record for this book is available from the British Library

ISBN 978-0-00-723014-3

Set in New Baskerville

Printed and bound in Great Britain by Clays Ltd, St Ives plc

For Theresa

Contents

There Leviathan,
Hugest of living creatures, on the deep
Stretch'd like a promontory sleeps or swims,
And seems a moving land; and at his gills
Draws in, and at his breath spouts out a sea.

John Milton, *Paradise Lost*, quoted in title page
to the first, English edition of *Moby-Dick*

Prologue

> For thou didst cast me into the deep,
> Into the heart of the seas,
> And the flood was round about me;
> All thy waves and billows passed over me.
>
> Jonah 2:3

Perhaps it is because I was nearly born underwater.

A day or so before my mother was due to give birth to me, she and my father visited Portsmouth's naval dockyard, where they were taken on a tour of a submarine. As she climbed down into its interior, my mother began to feel labour pains. For a moment, it seemed as though I was about to appear below the waterline; but it was back in our Victorian semi-detached house in Southampton, with its servants' bell-pulls still in place and its dark teak staircase turning on itself, that I was born.

I have always been afraid of deep water. Even bathtime had its terrors for me (although I was by no means a timid child) when I thought of the stories my mother told of her own childhood, and how my grandfather had painted a whale on the outside of their enamel bathtub. It was an image bound up in other childish fears and fascinations, ready to emerge out of the depths like the giant squid in the film of *Twenty Thousand Leagues Under the Sea,* with its bug-eyed Nautilus, Kirk Douglas's tousled blond locks and stripy T-shirt, and its futuristic divers walking the ocean floor as they might stroll along the beach.

I thought, too, of my favourite seaside toy – a grey plastic diver which dangled in the water by a thin red tube through which you blew to make it bob to the surface, trailing little silver bubbles – but which also reminded me of those nineteenth-century explorers enclosed in faceless helmets and rubberized overalls, their feet anchored by lead boots. And in my children's encyclopædia, I read about the pressurized bathysphere, an iron lung-like cell in which men descended to the Marianas Trench, where translucent angler fish lured their prey with luminous growths suspended in front of their gaping, devilish jaws. I was so scared of these monsters that I couldn't even touch the pages on which the pictures were printed, and had to turn them by their corners.

Southampton's municipal swimming baths, with their verdigris roof and glass windows, were a place of public exposure and weekly torture on our school trips there. Ordered to undress, revealing chicken flesh and, on older boys, dark sprouting hair, we shivered in ill-fitting trunks as we stood on wet tiles which, I was told, could harbour all sorts of disease.

Padding out into the echoing arena where weak winter sun threw mocking ripples on the ceiling, we lined up to plunge in the shallow end, ordered into the water by our PE master, a wiry-haired man with an imperious whistle on a cord around his neck.

Once in, we were told to hold the hand-rail and kick away with our feet. With my fingertips turning blue with the cold and my tenacious grip, I created enough white water to seem proportionate to my effort, although it was really an endeavour to disguise my ineptitude. Then we took a polystyrene float, crumbling at the edges like stale bread, and were instructed to launch ourselves across. The far side was as unattainable as Australia to me, and the reward for success – a piece of braid to sew on one's trunks – was a trophy I was as likely to win as an Olympic medal.

I never did learn to swim. The barked instructions, the fear of sinking to the tiled bottom along with the old sticking-plasters and hair-balls, combined to create an unconquerable anxiety. I somehow associated swimming not with pleasure, but with institutions, hospitals, conscription and war, with being ordered to do things I didn't want to do. At the beach I'd make my excuses when my friends ran into the sea, pretending I had a cold. Throughout my childhood and my teenage years, I lived with this disability; I even came to celebrate it, perversely, as a strength.

It was only later, living alone in London in my mid-twenties, that I decided to teach myself to swim. In the chilly East End pool, built between the wars, I discovered that the water could bear up my body. I realized what I had been missing: the buoyancy of myself. It was not a question of exercise: rather, it was the idea of going out of my depth, allowing something else to take account for my physical presence in the world; being part of it, and apart from it at the same time. In a way, it was a conscious reinvention, a means of confronting my fears.

For the poet Algernon Swinburne, the sea was a sensuous vice, one that he revealed in his only novel, *Lesbia Brandon*, set in his childhood home on the southern coast of the Isle of Wight, with its dramatic rocky cliffs overlooking the waters of the English Channel. In the book – not published until 1950, forty years after Swinburne's death – its young hero, Herbert, learns to love the water: 'all the sounds of the sea rang through him, all its airs and lights breathed and shone upon him: he felt land-sick when out of the sea's sight, and twice alive when hard by it.' He even dares the waves 'like a young sea-beast . . . pressed up against their soft fierce bosoms and fought for their sharp embraces; grappled with them as lover with lover'.

Swinburne, the son of an admiral, had a picturesque beach from which to swim; I grew up in a suburb on the other side of the Solent – a place of working docks and cranes and shipyards, close to which my father worked in a cable factory, testing huge insulated telecommunication lines which ran along the Atlantic sea-bed, as if tethering England to America. From my box bedroom at the back of the house I could hear the ships' horns on foggy mornings; at night, clanking dredgers gouged out a route for the huge liners and container ships that ply Southampton Water. Here, the sea represents commerce, rather than recreation. A port is a restless place, a place of transit, rather than a place in itself. Here, everything orientates itself towards the water – even the area in which I lived, Sholing, was a corruption of 'Shore Land' – yet at the same time the city seemed to ignore it, as if it and the element that is the reason for its existence were two entirely separate entities.

I think differently about the water now. Every day that I can, I swim in the sea. I feel claustrophobic if I am far from the water; summer and winter, I plan my time around the tides. Sitting on the shingly beach, I watch the ferries pass each other, briefly joining superstructures before they part again, caught between somewhere and nowhere. Pushing out into the same waters that so excited the red-haired poet and bore up his pale, freckled body, I lie on my back, on a level with the land, letting the waves wash over me like a quilt. Unencumbered, unobserved, in the warm waters of late August or in the icy rough seas of December, I am buoyed up, suspended, watching the world recede along with my clothes on the beach.

Sometimes something gelatinous will brush against my leg – one of the cuttlefish that are often cast up on the shore, their mottled flesh, hard parrot beaks and slimy tentacles rotting away to reveal the chalk-white bone below. Sometimes I'll feel a sharp sting after an encounter with an unseen jellyfish. Yet I still go out of my depth, where no one can find me, where terns dive and cormorants bob, and where I have no knowledge of what lies below. I dream of bodies underwater, veiled yet animate, like the drowned woman in the lake in *The Night of the Hunter*, or the shark I thought I once saw in a Cornish cove from the top of a cliff. The way the water both reveals and conceals still disturbs me. It is a deceptive and heartless lover.

> Consider the subtleness of the sea; how its most dreaded creatures glide under water, unapparent for the most part, and treacherously hidden beneath the loveliest tints of azure.
>
> Brit, *Moby-Dick*

Cities and civilizations rise and fall, but the sea is always the sea. 'We do not associate the idea of antiquity with the ocean, nor wonder how it looked a thousand years ago, as we do of the land, for it was equally wild and unfathomable always,' wrote the philosopher, Henry David Thoreau. 'The ocean is a wilderness reaching around the globe, wilder than a Bengal jungle, and fuller of monsters, washing the very wharves of our cities and the gardens of our sea-side residences.'

The sea is the greatest unknown, the last true wilderness, reaching over three-quarters of the earth. Its smallest organisms sustain us, providing every other breath of oxygen that we take. Its tides and shores determine our movements and our borders more than any treaty or government. Yet as we fly over its expanses, we think of it – if we think of it at all – merely as a distance to be overcome. In our arrogance, we consider that we have tamed the ocean, as much as we have conquered the land.

> . . . man has lost that sense of the full awfulness of the sea which aboriginally belongs to it . . . Yea, foolish mortals, Noah's flood is not yet subsided; two thirds of the world it yet covers.
>
> Brit, *Moby-Dick*

Once you have seen it, it is impossible to forget, just as if you never saw it, it would be impossible to describe. The sea is always in my head, the means by which I orientate myself to the earth – even in Red Cloud, Nebraska, where I once queued on a hot afternoon to swim in a public pool, a big blue hole in the middle of the Great Plains. It was as far from the ocean as I have ever been, but somehow a memory of it at the same time. The utter absence of the sea made its existence all the more potent.

To the careless, the water may seem the same from one day to the next, but under observation it becomes a continuous drama, made up of a million vignettes or grand gestures, played out at the edge of the shore or on the open ocean. It is a natural spectacle capable of rising dozens of feet into the air, or lying low like a glassy pond, so mirrored that it might not be there at all, seamlessly joining earth to sky. Surging and peaking, self-renewing and self-perpetuating, it can take away as easily as it gives. It is as punitive as it is generous. Sometimes it seems to be a living creature itself, an all-engulfing organism through which all the world exists, yet we see so little of it as we go about our daily lives: a glimpse from the car or a plane, the smallest fraction, even as we are infinitesimal in turn, mere grains of sand. And as I linger on the sea wall on my bike, looking out over the water, calm and grey on an autumn afternoon, it is even more improbable to imagine that its unspoken surface was once broken by giant creatures.

> The Whale and Grampus have been captured in Southampton Water, and on such rare occasions there have been of course the usual arrangements for sight-seers. Small shoals of Porpoises often visit the estuary; and the visitor from inland counties may be pleasingly surprised, as he walks the Quays and Platform, to see at a short distance from the shore many of these singular fish rolling and springing on the surface of the water, then disappearing, and rising again at another point to renew their awkward gambols.
>
> Philip Brannon, *The Picture of Southampton,* 1850

In the early 1970s we went on a family outing to Windsor Safari Park, where the star attraction was a killer whale. My youngest sister, even more enthusiastic about whales than I was, bought a small colour brochure somewhat apologetically entitled

Dolphins can be fascinating
at Windsor Safari Park.

On the front cover was a grinning Flipper; on the back was an advertisement for Embassy Regal cigarettes which, we were informed, were 'outstanding value'.

'You will be amused and delighted,' the booklet went on, by 'some fact and figures which might increase your knowledge, and enhance your enjoyment of their performance. You might also want to take some pictures of your own – take as many as you like!'

After shots of animals lolling at the pool like beauty contestants or leaping in the air like acrobats, a new player appeared in the programme:

'He is growing at the rate of 1 foot per year,' we read – a fact that raised inevitable consequences, even as we took in the oversized swimming pool in front of us – 'and at only four and a half years old he is 16 feet long, weighs one ton, and eats between 80 and 100 pounds of herring a day.'

> He was specially caught for Windsor Safari Park off the coast of North America in 1970 and was flown to London by Boeing 707 in a special crate which allowed him to be sprayed constantly

> with water, keeping him cool and fresh. Eventually, by lorry and crane, he arrived in the dolphin training pool, and after a short time was ready to commence his training programme.

Only later would I learn that captive whales decline to eat, and are force-fed until they do. I was more concerned with the spectacle about to appear before my eyes.

I don't remember how Ramu made his entrance (although my sisters do); but as he appeared, this sleek, powerful creature with his glossy black and white markings, it seemed as though his shiny skin had been bleached by the chlorine that kept the pool turquoise-blue; a pale, mocking imitation of the ocean which lay far away from his zoological prison.

The whale went through his routine, responding to his trainer's demands like a lap dog. When he leapt in the air and landed with a splash – soaking the thrilled ringside audience at this orca circus – it was as if he were beaten by his captivity, even as his proud dorsal fin flopped impotently over his back.

'Here in their pool at Windsor,' the brochure reassured us, the performers 'should survive for a great many more years than in the sea, to delight and entertain their visitors'. Within two years Ramu had grown too big for his tank. In 1976 he was sold to Seaworld in San Diego, where he was renamed Winston, sired four offspring, and died, ten years later, of heart failure – one of more than two hundred killer whales to perish in captivity in the last quarter of the twentieth century.

Back at home, I painted a picture of the orca in my journal, varnished and pristine on the page. But there were already other entries in my book, new passions. I forgot about whales, and thought about other things.

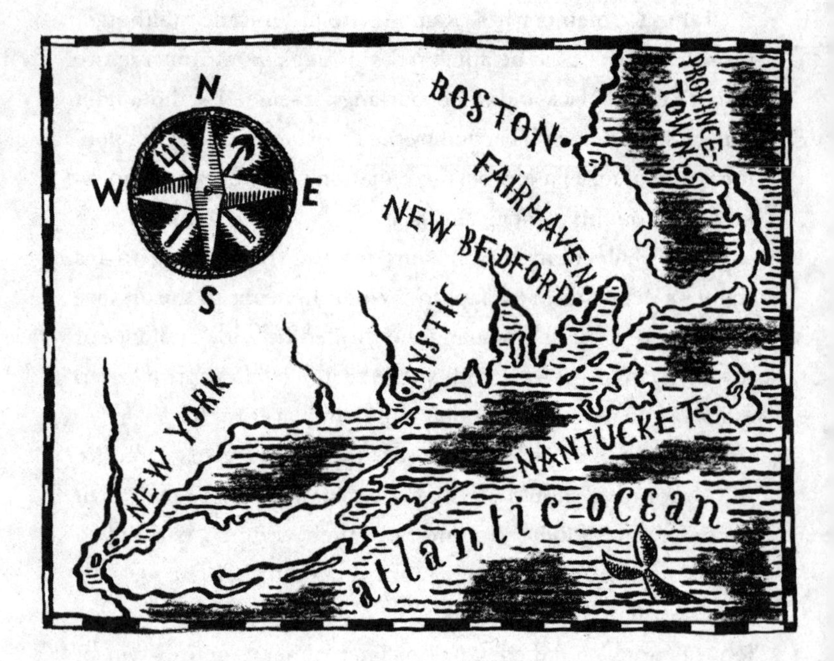
N
W
E
S
BOSTON
PROVINCE-TOWN
FAIRHAVEN
NEW BEDFORD
MYSTIC
NEW YORK
NANTUCKET
atlantic ocean

I

Soundings

> Chief among these motives was the overwhelming idea of the great whale himself. Such a portentous and mysterious monster roused all my curiosity.
>
> Loomings, *Moby-Dick*

It was my first visit to America. It was January, and I knew no one in New York. Freezing winds funnelled down the midtown canyons. Feeling homesick and lost, I took the subway as far as it would go. Outside the station at Coney Island, strange shapes stood in silhouette, skeletal versions of the Manhattan skyline I had left behind: a sinuous, hibernating roller coaster, and another instrument of amusement which looked like some giant gynæcological tool. I found my way to the aquarium and wandered through its empty interior, shuddering as I passed tanks filled with fish. There was something pathetic about this out-of-season place, a sense of abandonment blown in from the forlorn boardwalk and the suburban sea.

Let into the white walls was an observation window, thick enough to withstand tons of water. It reminded me of the portholes in Southampton's baths where children pressed their pasty

flesh to the glass; but this murky pane presented something entirely more spectral. Beckoning at the window, vertical and full length as if rising to greet me, was a beluga whale. It must have been twelve feet long, from its bulbous head to its stubby flukes; a huge ghostly baby fixing me with its stare.

As out of place as it seemed, this New York whale had an historical precedent. In 1861 Phineas T. Barnum had imported a pair of belugas to his American Museum on Broadway. Fished out of the waters off Labrador and brought south in hermetically sealed boxes lined with seaweed, the whales were twenty-three and eighteen feet long respectively. Their basement tank measured fifty-eight by twenty-five feet, but it was barely seven feet deep, and was filled with fresh water. In it they swam like lovers, although even their owner believed they would have only brief careers. 'Here is a real "sensation",' the *New York Tribune* marvelled, imagining that 'the enterprise of Mr. Barnum will not stop at white whales. It will embrace sperm whales and mermaids, and all strange things that swim or fly or crawl, until the Museum will become one vast microcosm of the animal creation.'

This fascination with the whale, like Philip Brannon's report from Southampton Water, was an expression of Victorian fashion, a characteristic marriage of ingenious science and human curiosity. In England, live whales were delivered to aquaria in Manchester and Blackpool (although one porpoise show was closed, for fear the flagrant activities of its performers should offend genteel dispositions), and in September 1877 a beluga whale arrived in Westminster, in the centre of the world's greatest city. The nine-foot, six-inch specimen had also been caught – along with ten others – off Labrador, where it had stranded at high tide and was netted by Zack Coup and his men. From there it began its long journey to London.

Taken in a narrow box by sloop to Montreal, the whale was put on a train to New York – a trip that took two weeks. The animal spent seven months at Coney Island's Summer Aquarium where 'he contracted his habit of swimming in a circle', before being taken out of its tank and put on a North German Lloyd steamship, the *Oder*, bound for Southampton. During the voyage, it was kept on deck in a rough wooden box lined with seaweed, and was wetted with salt water every three minutes. Despite such intensive care, the whale had already begun to live off its own blubber.

At Southampton the beluga was transferred to the South-Western Railway, travelling on an open truck to Waterloo Station and to its final home, an iron tank forty-four feet long, twenty feet wide, and six feet deep, at the Royal Aquarium, a grand gothic structure recently built opposite the Houses of Parliament. The whale waited as the tank took two hours to fill. 'He had been lying still in the box breathing once every 23 seconds. He flapped feebly with his tail when he felt them moving the box. He fell out of it sidelong into the water and went down to the bottom like lead.' The animal was allowed three hours of privacy before the public, 'in great numbers', were admitted to view it from a specially built grandstand.

The Times did not feel this was the right way to treat a whale. 'It is not likely he will live long in fresh water, although he comes up at intervals from ten to 100 seconds to breathe, and sometimes spouts the water up through the wide nostril which he has in the middle of his forehead. Noise or jarring caused by the workmen occasionally makes him stay beneath the water for two minutes at a time.' The beluga was fed live eels, but it was noted that its high dorsal ridge, 'which should be rounded with fat', stood up 'precipitously on his back'.

'Should he succumb to the unfavourable conditions of life in this city, no whalebone will be extracted from this monster,' the newspaper added. 'Nor is the white whale very rich in blubber. But his coat will make porpoise-skin boots.'

The Times's suspicions were correct, even if its assignation of gender was not. In what appeared to be delirious behaviour, the whale – which was in fact a female – swam up and down the tank rapidly, hitting its head on the wall. Then, 'having somewhat recovered, it again swam several times round the tank, again came into collision with the end of the tank, turned over, and died.'

Nor was the indignity over, for the body was taken out of the tank and exhibited to the public the next day. A plaster cast was made, and a necropsy performed by eminent naturalists and physicians. They discovered that far from starving, the whale had a full stomach – but also highly congested lungs. The fact that the animal had been kept on open deck on its way over the Atlantic, and, rather than keeping it alive, the regular dousing it had received, had resulted in rapid evaporation between soakings, causing it to catch cold.

The Westminster whale's public demise prompted correspondence from persons in high places. Bishop Claughton of St Albans, a poet in his own right, complained that it was 'the creature of which the Psalmist speaks as placed in its element by the Great Creator', and it was not man's right to take him out of it. William Flower of the Royal College of Surgeons – later to become the first director of the Natural History Museum – had attended the necropsy, and countered that the 'supposed marks of ill-usage' on its body 'were the consequences of the eels in the tanks having after its death nibbled the edges of its fins'. Professor Flower claimed the entire process was justified for 'the advantage to scientific and general knowledge to be gained'. But then, his own institution had benefited from the donation of the internal organs, which would 'make very interesting preparations'.

THE DEAD WHALE AT THE ROYAL AQUARIUM.

Back in New York, Barnum's whales met with their predicted fate. Victims of equally inappropriate conditions, like fairground fish brought home in plastic bags, they too had died within days – only to be replaced by successive specimens until a fire destroyed the museum in 1865. Futile attempts were made to rescue the last beluga, until a compassionate fireman smashed the tank with a hook, 'So the whale merely roasted to death instead of undergoing the distress of being poached.'

Faced with this modern captive on Coney Island, I felt a mixture of fascination and pity. It was as out of place as a tiger in a Manhattan apartment. The animal ought to have been swimming free in Arctic waters. Instead its pure white skin was soiled by its civic capture, as if the green algæ that covered the prismatic glass had contaminated it, too. It was struck dumb by the silence of that afternoon, and all the afternoons that stretched ahead. The beluga is the most vocal of all whales, known by sailors as the canary of the sea; here it was as caged as any tame songbird. As it hung there, this shrouded convict imprisoned for someone else's sins, I dared to touch it through the thick glass, as if something might pass between us. I waited for it to raise a flipper. But it didn't, so I turned away, unable to take its stare any longer.

After years living in London, the city had begun to press down on me. I sometimes felt as if all the sky were sea, and we citizens mere bottom-feeders, held down by its great pressure as we moved around the caverns and boulders of the streets. I lived on the borders of the City, within sight of the Docklands; over the years I watched the replicating skyscrapers rise up from the London clay like crystal stalagmites in a schoolboy's jam-jar experiment. At night I would dream that the tower block in which I lived was surrounded by water, inundated by the expected flood; that from my ninth-floor eyrie I could look down to see whales and sharks circling below. In other dreams, I saw a stone-walled harbour and a mass of marine animals caught within it, squirming and writhing to get out.

A place that had represented all my youthful aspirations now felt like a viral infection; and although, like a dose of malaria, I would never quite shake it, I was gradually, incrementally, leaving my old life behind. With the death of my father, and my mother living alone, I found myself spending more time back south. It was a kind of consolation, for grief and loss, for the severing of other emotional ties. I felt set adrift, anchorless – yet also a kind of convergence, a symmetry. It was the comfort of the old, but I saw it anew.

I replaced the treeless view from my ninth-floor flat with daily visits to the shore; the hard edges of the city with unconfined green and blue; stalking flea-bitten pigeons with black and white oystercatchers picking their way along the beach at low tide. My eyes stretched with the relief you feel when you look out over to the horizon from a train window, rather than onto the foreshortened visions of the street. Instead of superstitiously picking up pennies from the street, I combed the beach for stones with holes guaranteed to ward off witches, creating miniature avalanches as they piled up on my dressing table back home. And I stood looking out to sea, watching transatlantic ships sail by like Fitzgerald's boats borne back ceaselessly into the past, waiting for a future that might never come, like the man who fell to earth. As consoling as the water was, it sometimes served only to make me restless in my suburban exile.

Five years after my first visit to America, I took a train to Boston from New York's Penn Station. Having bought a map of New England from the bookstall, I began to trace my route along the coast. The name itself – a *New* England – seemed romantic, optimistic; both familiar and strange at the same time. The names on the map evoked the country I had left behind – Manchester, Norwich, Warwick – as Manhattan gave way to sharp sun and wide beaches and picnicking families, apparently unaware of the train hurtling past behind them. At the end of the line, I walked down to the harbour and boarded the ferry, watching Boston recede in a sequence of small islands, to the toll of a bell fixed to a buoy:

> fuller of dirges for the past, than of monitions for the future; and no one can give ear to it, without thinking of the sailors who sleep far beneath it at the bottom of the deep.

Ahead lay mile after nautical mile of sea. I did not know what to expect when I reached the other side, but as the boat docked, everyone else seemed to know where they were going. So I followed them, into Provincetown.

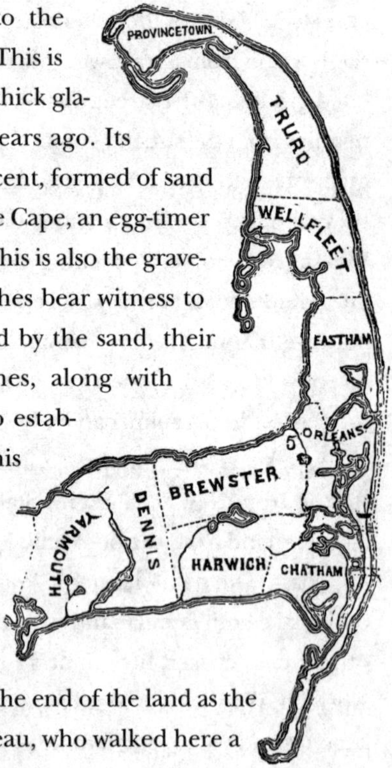

Cape Cod curls out into the Atlantic like a scorpion's tail. This is new land, carved out by mile-thick glaciers only fifteen thousand years ago. Its inner shores are still more recent, formed of sand carried from the far side of the Cape, an egg-timer adding even as it takes away. This is also the graveyard of the Atlantic. Its beaches bear witness to disaster: entire wrecks buried by the sand, their masts jutting from the dunes, along with human hands. Marconi, who established his radio station on this same shore, a forest of aerials among the marram grass, believed he could tune in to the voices of drowned men still hanging in the ether.

Cape Cod is not so much the end of the land as the beginning of the sea. To Thoreau, who walked here a hundred and fifty years ago, it was a place where 'everything seemed to be gently lapsing into futurity'. 'A man may stand there and put all America behind him,' he wrote; but this is where America began, too. Four centuries ago, the Pilgrim Fathers made first landfall on this sandy spit rather than at Plymouth Rock – just as they first left from Southampton, rather than Plymouth in Devon. In their search for utopia, the exiles found instead 'a hideous and desolate wilderness'. They had little idea that its native inhabitants had lived on the Cape for millennia.

After a month trudging through its sands, the Pilgrims rejected Cape Cod as fit only for fish and heathens. Provincetown became an outlaw colony beyond their Puritan influence, a reputation embodied by its nickname: Hell Town. Prey to piracy, war and revolution, by the end of the eighteenth century there were still only a handful of houses here. But soon this disputatious, barely legitimate port had entered its greatest prosperity – one that it owed to the whale.

The Pilgrims had regretted their lack of weaponry when they saw how many broad-backed, slow-moving whales lay in Cape Cod Bay. It was as if the animals were anchored to it. There were hundreds 'playing hard by us, of which in that place, if we had instruments and means to take them, we might have made a rich return'. Unlike the Indians who harvested whales for sustenance, Europeans sought profit in such animals, and had done so ever since the Basques had sailed to Labrador.

By the time the *Mayflower* set sail, other ships were leaving Dutch ports to carry out commercial whaling in the Arctic. Two of the crew of the *Mayflower* had whaled off Greenland, and reckoned they would have made £4,000 from the whales of Cape Cod Bay.

Indeed, it was the whales that had first prompted the Pilgrims to consider Provincetown as a site, and as Cotton Mather recorded, whale oil became the staple commodity of their colony. The *Mayflower* herself was pressed into service as a whaler, sailing over the bay from Plymouth.

Provincetown, too, took to whaling with aplomb. By 1737, twelve whale-ships were leaving the port, bound for the Davis Straits. By 1846, Provincetown was home to dozens of vessels. Families such as the Cooks, who owned eight houses in a row in the town's East End, could look out on their ships tied up in front of their properties much as modern cars are parked in driveways. The building that now houses a fashionable delicatessen was once the Cooks' chandlery. Close by stood the blacksmith's, forging harpoons and lances, while a blue plaque on another wall commemorates 'David C. Scull, the Ambergris King'. Later, the Azoreans and Portuguese came to work in the town's great salt cod trade. Their descendants still live here, incarnate in such names as Avellar, Costa, Oliveira and Motta, and in the annual Blessing of the Fleet, when their fishing boats are bedecked with flags and a dressed statue of St Peter is carried down to the harbour.

In the late nineteenth century other visitors came too, 'summer people' brought by steamers from Boston and New York, artists and writers among them. They were attracted by the clear light that bounces around the peninsula as from a photographer's reflecting shield, but also by its remoteness. Provincetown remained a tentative, if not dangerous place. The Portland gale of 1898 drowned five hundred people and demolished many wharves. Houses out on the sandy spit of Long Point, defeated by decades of storms, were floated wholesale across the bay on rafts of wrecking barrels to find shelter on calmer shores. As the radical journalist Mary Heaton Vorse wrote, 'Provincetowners have spent so much of their time on the sea in ships that they look upon houses as a sort of land ship or a species of house-boat and therefore not subject to the laws of houses.'

Gradually, reluctantly, the town was tamed. Drainage was installed, pavements laid, roads allowed access to what was, in effect, an island. 'Indeed, to an inlander, the Cape landscape is a constant mirage,' as Thoreau wrote. Its sands collect and drift as the town twists and turns on itself, leaving you never quite sure which way is south or which way is west. This is still a place apart, a fold-out on the map; not so much part of America as apart from it. In the summer it babbles with life, its one busy street teeming with day-tripping families and drag queens, before petering out at town limits once marked by a whale's jaw bone stuck in the ground, and now by Josh's garage and a straggle of beach huts from an Edward Hopper painting. And out on the ocean, the clamour diminishes like a dying chord, to be replaced by the rise and fall of the sea.

It wasn't until the day before I was due to leave Provincetown that I went on my first whale watch. I remember how cold it was as the boat left the bay, the land's warmth giving way to a chill sea breeze. As we sailed out of the harbour, our naturalist described the geography of Stellwagen Bank as it passed beneath us. He explained how fishermen had dredged up mastodon bones from the sea floor; how these were some of the most fertile waters on the planet; how they were crossed by the Atlantic's busiest shipping routes. On a chart behind him, he pointed out the animals we might see. I looked at their unlikely shapes on the pamphlet he had handed out. They seemed as unreal as the dinosaurs I'd memorized from my library books as a boy.

Then someone shouted,

Whale!

and in the mid-distance, a massive grey-black shape slid up out of the water and back down below. Before I knew it, there they were, off our bows, whales blowing noisily from their nostrils, rolling with the waves. Barely yards away a young humpback threw itself out of the water, showing off its white underbelly, ridged like some giant, rubbery shell. It was a jump-cut close-up of something impossible: a whale in flight.

Forgetting the children around me, I blurted out an inadvertent 'fuck!'. Other whales were throwing their tails in the air, slapping their flippers as though signalling to each other, or to us. As I watched, more and more animals appeared, as if summoned by some unseen circus master. I was amazed by the exuberant mastery of their own bodies, and the element in which they moved so elegantly. I envied them the fact that they were always swimming; that they were always free.

Every summer, humpbacks come to the Gulf of Maine. For six months they have fasted, and mated, in the warm but sterile waters of the Caribbean, suckling their calves with milk so rich it resembles cottage cheese, until it is time to make the annual pilgrimage north. It is the greatest migration undertaken by any mammal. Following routes of colonization first undertaken by their ancestors millions of years ago, navigating up to eight thousand miles of ocean via age-old and invisible signs, they arrive off the north-eastern seaboard, where the warm Gulf Stream meets the chill Labrador currents and stirs up nutrients from the ocean floor in a process called upwelling.

Here, in the grey-green waters, a vast food chain is set in motion. The whales fatten themselves on sand lances and herring, growing fat with the seasonal glut. And here, less than two hours' sail from one of America's great cities, these gigantic animals – 'the most gamesome and light-hearted of all the whales' – besport themselves, 'making more gay foam and white water generally than any other'. Even their hunters acknowledged this playfulness in their nickname for the humpback, the merry whale, although its scientific name is hardly less glamorous: *Megaptera novæangliæ*, big-winged New Englander, barnacled angel.

Launching fifty tons of blubber, flesh and bone into the air, the leviathan leaves its domain, its fifteen-foot flippers like gnarled wings, the tips of its tail, three times as wide as a man is long, barely in contact with the water.

Seen in the slow motion of recall – the after-image it leaves in your head – a breaching whale seems to be trying to escape its environment, the element that, even as it breaks the surface, is pulling it back down. No one really knows why whales leap. Almost every species does it – from the smallest dolphin to the greatest blue whale – in their own style: backward breaches, belly-flops, half-hearted lunges or full-blown somersaults. It may be that the animals are trying to dislodge parasites – the force is enough for breaching whales to slough off skin, convenient samples to be gathered for genetic tests. There is no knowing when they will breach, although when they do, they may do so repeatedly, often when the wind picks up, as if, like some cetacean Mary Poppins, a change in the weather summons their magical appearance. One scientist reasons that the gymnasts may find it 'more pleasurable or satisfying, or less painful, to slam the body on rough, rather than smooth, water'.

It seems likely that their aerobatics are an energetic means of communication – advertisements of physical power and presence, telling other whales, 'Here I am,' and 'Aren't I splendid?' But when you see a whale leap out of the water like a giant penguin, your first thought is that it looks *fun.* The fact that calves and young whales are more prone to breach reinforces this idea. The whales may be merely playing, like the boys who dive off Provincetown's Macmillan Wharf, placing implicit trust in their immortality as they hurl themselves from one medium to the other. Or perhaps they pity us for our enslavement to gravity, allowing us a glimpse of their true nature by rising out of the ocean to reveal their majesty.

Seeing whales in the wild seemed to turn me back into a boy. I remembered what it was that fascinated me about these outlandish animals: their sheer variety, their wildly differing shapes and sizes; a satisfying set to be collected like bubble-gum cards, a catalogue of complexity and colour: from the tiny harbour porpoise to the great rorquals – from the Scandinavian for reed or furrowed whale, a reference to their ridged bellies – and the mysterious sperm whale, a tiny model of which I found in my sister's toy box, still perched on its own plastic wave. It was as if the watery world I feared was restocked with friendly creatures, an international tribe of global roamers; as discrete and wide-ranging as birds, yet all of a type. This was what appealed to me: their completeness, as opposed to our separateness, for all that we are mammals together. They are a tidy whole; we are in disarray.

Cetaceans – from the Greek *ketos* for sea monster – fall neatly into two suborders. The toothed odontocetes – seventy-one species of porpoises, river and ocean dolphins, beaked whales, orcas and sperm whales – feed on fish and squid. The mysticetes or moustached whales – of which there are at least fourteen species – filter their diet of plankton and smaller fish through their baleen.

The bizarre nature of baleen seems to underline the otherness of the whale – one that begins in the womb. Although mysticete fœtuses have teeth buds, these are resorbed into their jaws before being born, to be replaced by sprouts of fibrous protein called keratin, the same material that furnishes humans with their fingernails. These long flat slats form pliable plates which line their gums in a great horseshoe shape, smooth edges outwards. They are continually growing, and are teased into fringes at their extremities by the constant play of the animal's tongue. Swallowing swimming pools of water – so greedily that they actually disarticulate their jaws to maximize their intake – baleen whales expand the ventral pleats in their bellies, then contract them to expel the surplus water and thereby catch their food in the bristles.

Toothed whales pursue their quarry through the ocean, fish by fish. Baleen whales are grazers and gulp mouthfuls at a time, from herring and sand eels to the tiny zooplankton which drift through the seas like animated dust. Here in the fertile waters of Cape Cod, it is the mysticetes that reign: from the elusive, relatively diminutive minke and the performing humpback, to the rotund right whale and the sleek fin whale – the second largest animal in the world, known as the greyhound of the sea, able to reach twenty knots or more.

After the blue whale, the finback, *Balænoptera physalus*, is also the loudest of any animal; and since sound travels further and faster through water, an American fin whale (if it cared about such things as nationalities) could be heard by its European counterpart on the other side of the Atlantic. Its mating call registers below the lowest level of human hearing; when it was first detected by scientists, they thought it was the noise of the ocean floor creaking. And in a few seconds, this immense creature – larger than any dinosaur – will pass beneath me. Lowering its broad, flattened snout, the whale dips below the keel in one imperceptible motion, as if powered by an invisible, silent motor.

> There you stand . . . while beneath you and between your legs, as it were, swim the hugest monsters of the sea, even as ships once sailed between the boots of the famous Colossus at old Rhodes.
>
> The Mast-Head, *Moby Dick*

In that one motion, my entire presence is undermined. I feel, rather than see, this eighty-foot animal swimming below. Knowing it is there tugs at my gut, and something inside makes me want to plunge in and dive with it to some unfathomable depth where no one would ever find us.

The finback completes its manœuvre, emerging on the larboard side to breathe; unlike humans, whales must make a conscious decision to respire, otherwise their dives would be impossible. With all the force of its massive lungs, it expels exhausted air with the pneumatic sound of a finger held over a bicycle pump. It is a profound exhalation, rather than a spout of sea water; a visible condensation, like human breath on a frosty morning.

From its organ valve nostrils, the whale shoots out one hundred gallons of air in a second, each cloudy discharge creating its own rainbow in the sun; then it repeats the process again and again, charging its body with oxygen until it is ready to dive once more, an act of internal transformation. Collapsing its lungs – a special mucus prevents the organs sticking together – and folding in its ribs along joints on the sides of the body, all remaining air is driven into 'dead spaces' within the whale's skull. This technique, and the lack of nitrogen in its bloodstream and air in its bones, prevents the animal from suffering the bends. More subtle than any submarine, the whale is a miracle of marine engineering.

With a last plosive *whoosh* as it fills its lungs, the finback shoots out a mixture of air and salt water and a little whale phlegm, its shiny blowholes closing in an airlock as it prepares to dive. The spume hits my face like a fishy atomizer. I have been breathed upon, and it feels like a baptism.

It is difficult not to address whales in romantic terms. I have seen grown men cry when they see their first whale. And while it is a mistake to anthropomorphize animals merely because they are big or small or cute or clever, it is only human to do so, because we are human, and they are not. It is sometimes the only way we can come to an understanding of them.

Nothing else represents life on such a scale. Seeing a whale is not like seeing a sparrow in a city tree, or a cat crossing the street. It is not even like seeing a giraffe, dawdling on the African veldt, batting its glamorous eyes in the dust. Whales exist beyond the normal, beyond what we expect to see in our daily lives. They are not so much animal as geographical; if they did not move, it would be difficult to believe they were alive at all. In their size –

their very construction – they are antidotes to our lives lived in uncompromising cities. Perhaps that's why I was so affected by seeing them at this point in my life: I was ready to witness whales, to believe in them. I had come looking for something, and I had found it.

Here was an animal close to me as a living creature – one that shared my heart and lungs, my mammalian qualities – but which at the same time was possessed of a supernatural physicality. Whales are visible markers of the ocean life we cannot see; without them, the sea might as well be empty for all we know. Yet they are entirely mutable, dreamlike because they exist in another world, because they look like we feel as we float in our dreams. Perhaps, without our projections, they would be merely another species, another of God's creation (although, of course, some might say that's just another projection in itself). Nevertheless, we imbue whales with the improbability of their continued existence, and ours. We are terrestrial, earthbound, dependent on limited senses. Whales defy gravity, occupy other dimensions; they live in a medium that would overwhelm us, and which far exceeds our own earthly sway. They are Linnæan-classified aliens following invisible magnetic fields, seeing through sound and hearing through their bodies, moving through a world we know nothing about. They are animals before the Fall, innocent of sin.

But they also have bad breath, and shit reddish water. They eat day and night without discretion. They are super-sized animals, 'charismatic megafauna' in the zoologists' dismissive phrase. They cannot, like the old joke, be weighed at a whale weigh-station, although they once were placed on scales in pieces, like legs of lamb. Out of their element, they collapse under their own weight, lacking limbs to support themselves, pathetically incapable of self-preservation despite, or because of, their great size. (One soon runs out of superlatives when writing about whales.) For all their physical reality, they cannot be encompassed, or even easily described. We may stand around in awe and pick apart their carcases, but in the end all we are left with to show for our curiosity are bones which give little clue to the true shape of their living owners.

Whales existed before man, but they have been known to us only for two or three generations: until the invention of underwater photography, we hardly knew what they looked like. It was only after we had seen the Earth from orbiting spaceships that the first free-swimming whale was photographed underwater. The first underwater film of sperm whales, off the coast of Sri Lanka, was not taken until 1984; our images of these huge placid creatures moving gracefully and silently through the ocean are more recent than the use of personal computers. We knew what the world looked like before we knew what the whale looked like. Even now there are beaked whales, or ziphiids, known only from bones washed up on remote beaches – esoteric, deep-sea animals with strange markings which biologists have never seen alive or dead, so little studied that their status is 'data deficient'. New cetaceans are still being identified in the twenty-first century, and we would do well to remember that the world harbours animals bigger than ourselves, which we have yet to see; that not everything is catalogued and claimed and digitalized. That in the oceans great whales swim unnamed by man.

In December 2004, the *New York Times* reported on the publication of an obscure scientific paper. *Twelve Years of Tracking 52-Hz Whale Calls From a Unique Source in the North Pacific* was the result of research on a whale cruising from California to the Aleutian Islands off Alaska, 'calling out with a voice unlike any other whale's, and getting no response'.

'The call, possibly a mating signal, suggests that the animal lives in total, and undesired, isolation.' The sound had been tracked for more than a decade, and in that time its timbre deepened, suggesting that the whale was still maturing. One scientist thought it might be 'miswired, broadcasting on the wrong frequency but listening on the right one'; another considered that the caller could be the miscegenic result of a liaison between a blue whale and another species, 'and hence truly alone of its kind'.

Such stories seem to tug at our hearts because we cannot help but invest emotion in these paradoxical animals. They feed on the tiniest organisms – whales have to be big to swallow such huge quantities – yet they need to eat large amounts to sustain their size. Humpbacks, for instance, eat a ton of fish a day, mostly sand eels which, with their salt-excreting glands, are full of fresh water and therefore sate the animals' thirst. Whales might live in the world's great bodies of water, but they can never drink.

Delicately attuned to their surroundings, whales announce their presence in sonar pulses; seeing in sound, they diagnose the condition of a world from which we are insulated by our ignorance. As products of a different branch of evolutionary selection, they appear to have arrived at a superior way of being. The open ocean, without barriers and with a ready supply of food, is an excellent medium for the evolution of such huge, long-lived and intelligent animals; an environment in which communication and socializing take the place of material culture. Theirs is a landless race, free from mortgages and fossil fuel, unconstrained by borders or want, content merely to sing and sleep and eat and die.

It has taken us almost all our existence to come close to the true nature of the whale; only in the last few decades have we come to realize what the whale might be. In the long view of history, it will seem a remarkable turn-around: that a century that began by actively hunting whales ended by passively watching them. Animals, too, have a history – although one we can know only a tiny part of – and while modern science has demystified the whale whilst revealing its true wonders, our attitudes to whales also changed when we came to see them close-up. When, in effect, they became mediated, in photographs, on film, on television, part of our public discourse.

For the modern world, the whale is a symbol of innocence in an age of threat. It is an animal out of Genesis, a 'myth of the fifth morning', in Mary Oliver's poem, both childlike and reproving. History, on the other hand, saw peril in the great fish that swallowed Jonah, or on which Sinbad found himself, a gigantic whale 'on whose back the sands have settled and trees have grown since the world was young!' The ancient writer Lucian told of a whale one hundred and fifty miles long in which was contained an entire nation and men who believed themselves to be dead, years after they were first engulfed. The beast that attacked Andromeda, and which was slain by Perseus, was believed to be a whale. Cetus was sent by Poseidon to consume the young of Ethiopia, only to be turned into a huge rock when it looked at the Medusa – a celestial myth re-enacted each autumn as the whale constellation rears over the southern horizon.

Although D.H. Lawrence would declare that 'Jesus, the Redeemer, was Cetus, Leviathan. And all the Christians all his little fishes', to the Christian era, the whale was the very shape of the Beast of Revelation. In the sixteenth century the metaphysical poet John Donne wrote of a monstrous animal,

His ribs are pillars, and his high arch'd roofe
Of barke that blunts best steele, is thunder-proofe

while a continent away in the New World, North-Western American Indians believed that the giant waves that carried away their villages were the backwash of battles between thunderbirds and whales. In the Hindu version of the flood, Vishnu assumes his first avatar in the shape of a great fish with a horn and tows Manu and his ark to safety, and followers of Islam contend that of the ten animals that will enter paradise, one is the whale that swallowed Jonah. Overwhelmingly, however, the modern whale exists in one great image, the looming shape of its most famous incarnation: Moby Dick.

> And the angel of the LORD said to her, 'Behold, you are with child, and shall bear a son; you shall call his name Ishmael; because the LORD has given heed to your affliction. He shall be a wild ass of a man, his hand against every man and every man's hand against him; and he shall dwell over against all his kinsmen.'
>
> Genesis 16:11–12

Like many people, I found the densely written chapters of Herman Melville's book difficult to read. I was defeated by its size and scale, by its ambition. It was as incomprehensible as the whale itself. Over the years I'd pick up the book, become engrossed, only for my attention to wander. But after my first visit to New England, I looked at it again; just as I was ready to see whales, I was ready to read *Moby-Dick*.

Perhaps it was the solace I'd found in reading *Billy Budd, Sailor, & Other Stories* during the endless hours of a transatlantic flight when, despite the darkened cabin and everyone else around me cocooned like larvæ in thin airline blankets, my own eyes resolutely refused to remain shut. The yellowing pages of a 1970s Penguin edition – bought when I was at college in London, studying English literature – seemed somehow consoling with their tales of travel in less constrained times, especially the elegiac story of the Handsome Sailor, a boy fated to die for sins not his own. Or perhaps it was the enigma of the author himself that intrigued me, a man who lived through an American century which he foretold, yet who died forgotten at its end.

Published in the middle of that century, in 1851 – four years after *Wuthering Heights*, the only novel to rival its mysterious narrative power – *Moby-Dick* drew on Melville's own experiences of a whaling voyage ten years before. The book begins with a startling, modern abruptness, launching itself on the reader like a rushing wave with the most evocative opening line of any work of fiction:

Call me Ishmael.

From this deliberately equivocal declaration – is this our hero's real name, or merely a convenient disguise? – and its biblical overtones, we follow the rootless young man from Manhattan, where he has become so tired of life that he feels murderous, even suicidal, to his chosen refuge, the sea. From New

Bedford, Ishmael sails around the world in pursuit of whales. His intentions are both poetic and prosaic: 'I always go to sea as a sailor,' he says wryly, 'because they made a point of paying me for my trouble, whereas they never pay a passenger a single penny that I ever heard of.'

For his half-demented, peg-legged captain, Ahab, however, the voyage of the *Pequod* is an extended act of vengeance against a monstrous sperm whale: a terrifying, toothed creature of the deep ocean, rather than the placid baleen whale of coastal waters. This is the beast that dismasted Ahab, and which will in time take the rest of him, too. Even in this new industrial century, man still feared the elements of nature; and as the wild Yorkshire heath is itself a character in Emily Brontë's book, so for Melville the whale was the unholy instrument of fate. Not for nothing is Ahab warned by the mad prophet Gabriel of the passing ship, *Jeroboam*, that the White Whale is 'the Shaker God incarnate'. Jonah was saved by the whale for God's work; Ahab is destroyed by the devil's. Only Ishmael survives as 'another orphan', an emblem of martyrdom and rebirth, for a man must lose his life to save it.

Moby-Dick surpasses all other books because it is utterly unlike any other. It stands outside itself from the start, with its introductory list of historical quotations pertaining to the whale, as gathered by Ishmael's 'sub-sub-librarian'; and from there it moves through eccentric taxonomical descriptions as Melville attempts to capture his subject even as his hunters sought to harpoon it. Sidestepping his own narrative even as he delivers it, Ishmael almost wilfully and continually interrupts the reader with diversions and digressions, pulling him aside to address him with hell-fire sermons or musical interludes, with anatomical allegories or sensual dissertations on spermaceti oil.

In chapter after chapter, Melville teases out new legends to encircle the world and the whale. He creates a new family of men bent in pursuit of the whale, and a new kind of existence, culled from the lives he himself witnessed. Out of the oily, grimy labour of whaling, he forges a sterling heroism. In doing so, he melds his experience at sea with his dark view of the world and the nature of good and evil itself, seeing the future of his nation through his immaculate yet blasphemous creation, as if the whale were an American Sibyl of the new age.

Now, as I came to it again, I saw that *Moby-Dick* is a book made mythic by the whale, as much as it made a myth of the whale in turn. It is the literary mechanism by which we see the whale, the default evocation of anything whalish – from newspaper cartoons and children's books to fish and chip shops and porn stars. Few could have predicted such an outcome for this eccentric work, least of all its author. *Moby-Dick* failed to sell out its first edition, and was almost entirely ignored in Melville's lifetime. It took a new century for its qualities to be appreciated. In 1921 Viola Meynell declared that 'to read it and absorb it is the crown of one's reading life', and wrote of its author, 'His fame may still be restricted, but it is intense, for to know him is to be partly made of him for ever.' (She also noted that J.M. Barrie invented Captain Hook out of Ahab, and his pursuant, time-ticking crocodile from the White Whale.) Two years later, in his extraordinary collection of rhetorical essays, D.H. Lawrence wrote: 'He was a futurist long before futurism found paint . . . a mystic and an idealist', author of 'one of the strangest and most wonderful books in the world, closing up its mystery and its tortured symbolism'.

Moby-Dick became the great American novel retrospectively. It also became a kind of bible, a book to be read two pages at a time, a transcendental text. Each time I read it, it is as if I am reading it for the first time. I study my tiny edition as I ride on the Tube, as intently as the veiled woman next to me reads her Koran. Every day I am reminded that it is part of our collective imagination: from newspaper leaders that evoke Ahab in the pursuit of the war on terror, to the ubiquitous chain of coffee-shops named after the *Pequod*'s first mate, Starbuck, where customers sip to a soundtrack generated by a great-nephew of the author, Richard Melville Hall, better known as Moby.

Melville's White Whale is far from the comforting anthropomorphism of the smiling dolphin and the performing orca, from *Flipper* to *Free Willy*, or the singing humpback and the 'Save the Whale' campaign – all carriers, in their own way, of our own guilt. Rather, Moby Dick's ominous shape and uncanny pallor, as seen through Ahab's eyes, represents the Leviathan of the Apocalypse, an avenging angel with a crooked jaw, hung with harpoons from the futile attempts of other hunters. This whale might as well be a dragon as a real animal, with Ahab as his would-be slayer.

The age of whaling brought man into close contact with these animals – never closer, before or since. The whale represented money, food, livelihood, trade. But it also meant something darker, more metaphysical, by virtue of the fact that men risked their lives to hunt it. The whale was the future, the present and the past, all in one; the destiny of man as much as the destiny of another species. It offered dominion, wealth and power, even as it represented death and disaster, as men met the monster eye to eye, flimsy boat to sinewy flukes, and often died in the process. More than anyone has realized, perhaps, the modern world was built upon the whale. What was at stake was the future of civilization, in the most brutal meeting of man and nature since history began. And as the animals paid for the encounter in their near extinction, so we must ask what price we paid in our souls. How have we moved so far from one notion of the whale to the other, in such a short space of time?

When I close my eyes, I see those massive animals swimming in and out of my vision, into the blue-black below; the same creatures that came to obsess Melville's ambiguous narrator, 'and in the wild conceits that swayed me to my purpose, two and two there floated into my inmost soul, endless processions of the whale'. On my own uncertain journey, I sought to discover why I too felt haunted by the whale, by the forlorn expression on the beluga's face, by the orca's impotent fin, by the insistent images in my head. Like Ishmael, I was drawn back to the sea; wary of what lay below, yet forever intrigued by it, too.

II

The Passage Out

> There now is your insular city of the Manhattoes, belted round by wharves as Indian isles by coral reefs – commerce surrounds it with her surf. Right and left, the streets take you waterward. Its extreme down-town is the Battery, where that noble mole is washed by waves, and cooled by breezes, which a few hours previous were out of sight of land. Look at the crowds of water-gazers there.
>
> Circumambulate the city of a dreamy Sabbath afternoon . . . What do you see? – Posted like silent sentinels around the town, stand thousands upon thousands of mortal men fixed in ocean reveries . . . Nothing will content them but the extremest limit of the land . . . Tell me, does the magnetic virtue of the needles of the compasses of all those ships attract them thither?
>
> Loomings, *Moby-Dick*

Nowadays Pearl Street is covered with asphalt, but once it was strewn with oyster shells, like the glistening white paths you can still see on Cape Cod. On 1 August 1819, when Herman Melville was born here, this thoroughfare marked the lower limits of Manhattan. And if it is hard now to imagine what New York looked like without its towers, rising to the sky in an insatiable search for space, then it was a notion familiar to Melville, for the city changed utterly within his own lifetime.

In 1819 much of Manhattan was still farmland; Central Park had yet to be born out of the common ground where freed slaves and the last Native Americans lived. Most New Yorkers were British or Dutch by descent; this was not the polyglot city it would become by the century's end. The shallows in which the oysters grew were yet to be clawed back from the sea, and at the end of Pearl Street was the Battery, a promenade where citizens could take the sea air. Its Castle Clinton was still an island, although it would later become the home of the New York Aquarium where, in 1913, Charles H. Townsend exhibited a live porpoise.

The house in which Melville was born was demolished long ago. Set into a wall nearby is a memorial bust of the author, covered by perspex like a square porthole and overshadowed by an office block. Across the road, the river ferries spill out their early morning commuters from Jersey, in the shadow of the moored, anachronistic masts of South Street Seaport.

The sun shines through the cables of Brooklyn Bridge; a down-and-out stirs from a riverside bench. This is still a fluid place, accustomed to reshaping itself in its own image and leaving its history behind. Yet the past remains imprinted in these streets, and in the memory of the people who once walked them.

They were what we would call middle class. Herman's father, Allan Melvill – the 'e' was added later as a claim on their noble Scottish ancestry – was an importer of fancy goods. A dandified figure with his brushed-forward hair, he had made many trips to Europe, bringing back French antiques and engravings over which his children pored on a Saturday afternoon. 'Above all there was a picture of a great whale, as big as a ship, stuck full of harpoons, and three boats sailing after it as fast as they could fly.' Such images left his young son with 'a vague prophetic thought, that I was fated, one day or other, to be a great voyager'.

On both sides Melville sprang from heroes. His paternal grandfather, Major Thomas Melvill, was one of the 'Indian' raiders who tipped tea into Boston harbour in protest at British taxes; the family kept a phial of the tea leaves in his honour. His other grandfather, General Peter Gansevoort, after whom his brother was named, had held Fort Stanwix in the 1777 siege against the British and the Indians; Herman would call his own son Stanwix in memory of this famous victory. The sea was in the family blood, too. One uncle, Captain John D'Wolf II, had sailed from the Kamchatka Peninsula and onto the back of a whale. 'It was like striking a rock, and brought us to a complete standstill,' he recorded. 'The monster soon showed himself, gave a spout, "kicked" his flukes and went down. He did not appear to be hurt, nor were we hurt, but most confoundedly frightened.' A fine, handsome man with white hair and a florid face, D'Wolf was the first captain young Herman had ever met. He was later lost at sea.

With their growing family, the Melvills moved steadily uptown in a succession of grander houses until they reached 675 Broadway – a neighbourhood known as Bond Street whose gentility has long since been swept away by the waves of commerce and cheap denim. Here Herman and his brothers and sisters were taught by a governess, although a bout of scarlet fever damaged his eyesight and made it difficult to read. Life seemed stable enough, but in 1830 their father was declared bankrupt. The family were forced to move to Albany, the state capital up the Hudson River. Two years later, aged forty-eight, Allan died in a maniacal fever, leaving his wife Maria with only debt and eight children in her care.

At the most formative point in his life, twelve-year-old Herman was cast adrift, losing all sense of security when he most needed it. He would later claim that his mother, a strict Calvinist, hated him. He left school to work in a bank, but could not settle, and after a spell teaching and working on his uncle's farm, he went west, hoping to become a surveyor on one of the new canals that were opening up the American interior. He got as far as the frontier, St Louis, Missouri, before returning to New York, where he was declined employment as a lawyer's clerk because his handwriting was so bad. 'There is no misanthrope like a boy disappointed, and such was I, with the warm soul of me flogged out by adversity.' Rejected by the land, the young man sought a new life at sea.

On 5 June 1839 the *St Lawrence* sailed from New York with a cargo of cotton destined for Lancashire mills. Also on board was the nineteen-year-old Herman Melville. He was an outsider, abused by the crew for his middle-class manners, his dandified clothes, and his ignorance of shipboard life, 'so that at last I found myself a sort of Ishmael . . . without a single friend or companion'. He found consolation in the ocean, which swelled unaccountably as if possessed of a mind of its own. Once, in a Newfoundland fog, he heard the sound of sighing and sobbing which sent him to the side of the ship. There he saw 'four or five long, black snaky-looking shapes, only a few inches out of the water'. These were not the monstrous whales of his father's engravings, no 'regular krakens, that . . . inundated continents when they descended to feed!' They even made him wonder if the story of Jonah could be true.

The sights of Liverpool, the second city of the Empire, amazed the young man. He saw a Floating Chapel converted from an old sloop-of-war, with a steeple instead of a mast, and a balcony built like a pulpit. Here William Scoresby, once one of England's greatest whalers and now a man of the cloth, preached. There were scenes of shocking poverty, too. One young man silently exhibited a placard depicting himself 'caught in the machinery of some factory, and whirled about among spindles and cogs, with his limbs mangled and bloody'. And in an even more horrific image, a nameless shape moaned at the bottom of some cellar steps: a destitute mother with two skeletal children on either side and a baby in her arms. 'Its face was dazzlingly white, even in its squalor; but the closed eyes looked like balls of indigo. It must have been dead some hours.'

On 30 September Melville returned to New York on the *St Lawrence*, only to find nothing had changed but himself. He had made no money, and had to go back to teaching to support his widowed mother and his four sisters. But he had known life at sea, and within a year he would leave on an even more ambitious voyage – from the Whaling City itself.

> The transition is a keen one, I assure you, from a schoolmaster to a sailor . . .
>
> Loomings, *Moby-Dick*

In the second chapter of *Moby-Dick,* Ishmael arrives in New Bedford on a snowy Saturday night, only to discover that he has to wait two days until the next packet sails for Nantucket, where he intends to join his ship. Searching the shore-huddled town for a cheap bed for the night, he finds the Spouter Inn, its timbered interior hung with 'horrifying implements' and murky paintings of impenetrable sea-scenes. Here he is told by the landlord that he must bunk with a harpooneer.

There was nothing so unusual in that; Abraham Lincoln himself often shared his bed with a travelling companion. But Ishmael is aghast to find that his room-mate is a six-foot savage with a tattooed face. 'Such a face! It was of a dark, purplised, yellow color, here and there stuck over with large, blackish looking squares.' And as Queequeg puts aside the mummified head he has been trying to sell in town and undresses by candlelight, Ishmael realizes with horror that the cannibal's entire body is tattooed, too.

This is the man with whom he is expected to spend the night. After some hullabaloo, however, the white American lies down with the blue-stained Polynesian, and in the morning, Ishmael awakes to find Queequeg's arm tight around his body 'in the most loving and affectionate manner. You had almost thought I had been his wife.' But as he lies there, unable to move, the young man is taken back to a childhood memory, of darkness, claustrophobia and terror.

It was midsummer's day. For some minor misdemeanour, the infant Ishmael was sent to bed early. He endured the awful punishment of confinement while the world went on around him,

outside his bedroom. Coaches passed by, other children played. The sun shone brightly on the longest day of the year, defying his attempts to kill time.

Eventually, he fell into 'a troubled nightmare of doze', from which he awoke with his arm dangling down beside the bed – only to find another hand clasped in his own. 'For what seemed ages piled on ages, I lay there, frozen with the most awful fears, not daring to drag away my hand.' As he fell asleep again, the sensation left him; yet he could never reconcile the strange, half-waking, half-sleeping encounter he had had with 'the nameless, unimaginable, silent form' that had gripped his hand.

Lying there on that frosty December dawn in New Bedford, imprisoned by his bed-mate, Ishmael could barely distinguish Queequeg's arm from the counterpane. Both were so heavily patterned that they seemed to blend one into the other: 'this arm of his tattooed all over with an interminable Cretan labyrinth'; the patchwork cover with its 'odd little parti-coloured squares and triangles'. Far from being terrified, Ishmael is comforted by the sensation, secure in the giant man's embrace, as if he himself might become patterned all over, too. That night he becomes Queequeg's 'bosom friend'; the two would die for each other. Such is Ishmael's rebellion against the normal world, that he should so intimately identify with so pagan a figure.

These scenes, part nightmare and part romance, are some of the most memorable in Victorian literature, so vividly written one might almost believe the author had experienced them himself. But when he arrived at the wintry port in the Christmas of 1840, Melville stayed on the opposite side of the river, at Fairhaven. He was accompanied by Gansevoort, who bought his younger brother the items he needed: an oilskin suit, a red flannel shirt, duck trousers; a straw tick, pillow and blankets; a sheath knife and fork, a tin spoon and plate; a sewing kit, soap, razor, ditty bag; and a sea chest in which to store them.

30 December 1840

LIST OF PERSONS

COMPRISING THE CREW OF THE SHIP
ACUSHNET OF FAIRHAVEN

Whereof the Master, Valentine Pease, bound for Pacific Ocean

NAMES	PLACES OF BIRTH	PLACES OF RESIDENCE	OF WHAT COUNTRY CITIZENS OR SUBJECTS
Herman Melville	Fairhaven	New York	US

DESCRIPTION OF THEIR PERSONS

AGE	HEIGHT FEET	HEIGHT INCHES	COMPLEXION	HAIR
21	5	9½	Dark	Brown

Of the twenty-six men about to sail on the *Acushnet*, all had a share or lay in her future – fractions as eloquent as any amount of gold braid. Captain Pease, master and part owner, claimed 1/12 of all profits; the first officer, Frederic Raymond of Nantucket, 1/25. As a foremost hand, Melville's lay was 1/75; while lowly Carlos Green of New York – a true greenhand – could expect just 1/190. For some, even that was welcome, not least William Maiden, the cook, and deckhands Thomas Johnson and Enoch Read, whose complexions were recorded as black or mulatto. They had ever laboured under a master; now they had signed away their lives to the whale.

The *Acushnet* was fresh off the production line; at the peak of the whaling boom, new whale-ships were said to be built by the mile, 'chopped off the line, like sausages'. Others were converted liners or packets. 'Thus the ship that once carried over gay parties of ladies and gentlemen, as tourists, to Liverpool or London, now carries a crew of harpooneers round Cape Horn into the Pacific'. Quarterdecks where the gentry once took the sea air now reeked of whale oil. 'Plump of hull and long of spar', the *Acushnet* was 104 feet long, 27 feet wide and 13 feet deep. Named after the river on which she was launched, she towered over the wharf at Fairhaven, her web of rigging and tall masts statements of industry and fortitude. Unlike her alter ego, she lacked bulwarks studded with whale teeth or a tiller fashioned from a whale's jaw, embellishments that gave Ahab's *Pequod* the air of a 'cannibal of a craft, tricking herself forth in the chased bones of her enemies'. The *Acushnet* had her own disguise: false gunports painted on her side to ward off attacks from pirates or savages.

She was owned by a syndicate of eighteen men, among them the agent, Melvin O. Bradford and his brother, Marlboro

Bradford, both Quakers. Their captain, Valentine Pease Junior, was forty-three years old, a tall, stern, bewhiskered and sometimes profane man, not overwhelmingly blessed with luck. On his first command, the *Houqua,* his first mate, Edward C. Starbuck, had been discharged in Tahiti 'under conditions curious and not fully explained'. Seven men drowned, two others died when their boat was stove in by a whale, and eleven crew deserted, leaving only three original members to return and claim their lays.

This was not an unusual story. Of her original crew of twenty-six, only eleven would return on the *Acushnet.* The rest deserted or were discharged, discouraged by long and inhospitable voyages and strictures enforced by omnipotent captains. Contracts stated that men were not to leave the ship until her hold was full of oil, and that they must adhere 'to the good order, effectual government, health and moral habits' expected of them. 'Criminal intercourse' with women would be punished by the forfeit of five days' pay; 'intemperance and licentiousness' earned similar penalties, if not the lash. To add insult to injury, wear and tear meant that they had to buy new clothes from overpriced onboard supplies. When the debt was deducted from their share of the ship's profits, they were often left with nothing, or even found themselves owing money for their trouble. Given such conditions, it was hardly surprising that men jumped ship. In fact, two of the *Acushnet*'s crew had deserted even before she sailed. They had not signed up to be slaves, after all.

There are some things a place will not tell you, as if it conspires with its past. To look at it now, you would not guess that New Bedford was once the richest city in America. This now incongruous town – at least, to anyone who has not been there – was the capital of a new economy, one that reached out across the world; the bustling industrial centre of a republic founded on the backs of whales.

New Bedford's roots lay in its sheltered harbour and good connections with the rest of New England, but, above all, strong ties with the Quakers of Nantucket – who had perfected the art of whaling in the early eighteenth century – con-tributed to the port's unprecedented success. One of those Quakers, Joseph Rotch, developed New Bedford in the years following the Revolution. By the 1840s, when Melville arrived, the port had grown rich – more so since it was linked by a bridge with Fairhaven, its twin on the other side of the river.

Route 6, the highway once known as the King's road and which runs all the way to the tip of Cape Cod, still crosses the Acushnet by a nineteenth-century turntable bridge, a Meccano construction that pivots to permit more important traffic to pass. Here vessels still have precedence over cars. This is a working port. It smells of diesel and fish, and there are ships at the end of its streets. It is also a designated national park, not of rolling hills or woods, but of thirteen city blocks, all devoted to a memory.

New Bedford – The Whaling City

Set next to the modern freeway, on a huge, block-like plant for refrigerating fish, is a giant mural of air-brushed whales swimming serenely in a turquoise-blue sea. The whale is imprinted on New Bedford: even the licence plates of the cars that drive through it are embossed with the sperm whale, the state animal of neighbouring Connecticut.

In front of the Free Public Library is an outsize statue on a granite block. It resembles a war memorial, but it was set in place in 1913, and carved with a succinct epithet –

A DEAD WHALE OR A STOVE BOAT.

– a simple enough equation. Despite his square jaw and Aryan looks, there is something tribal about the idealized, muscular whaler balanced on a disconnected prow; he might almost be a Plains Indian. His spear is aimed at one inexorable point: we are the whale; this was the first human it saw, and the last.

Let his monument stand, with his harpoon in hand.
Sturdy son of the sea who dragged wealth to the land.

Modern New Bedford lives on in the shadow of such monuments. Brooks Pharmacy sells garish post-cards of the Whaling City. Visitors can 'Catch the Whale', a downtown shuttle bus, or buy T-shirts from the Black Whale shop. Around the corner, the dark interior of Carter's menswear store, est.1947, is piled high with workwear and fishermen's caps for modern Ishmaels. The young assistants nod to their few customers on a Saturday morning, preferring to get on with talking about their Friday nights. Tomorrow, the church steeple over the way will summon sailors to the Lord, along with the sleepy guests from the Spouter Inn.

> In this same New Bedford there stands a Whaleman's Chapel, and few are the moody fishermen, shortly bound for the Indian Ocean or Pacific, who fail to make a Sunday visit to the spot. I am sure that I did not.
>
> The Chapel, *Moby-Dick*

At the entrance of the Seaman's Bethel – which, with its clap-board and its square tower, resembles a ship sailing over the brow of Johnny Cake Hill – a veteran from the mission next door shows me inside, then steps out for a smoke, leaving me to wander around alone. The dark hallway opens into an airy space lined with box pews and white marble slabs set into the wall, each a witness to past mourning, 'as if each silent grief were insular and incommunicable'.

In Memory of
CAPT. WM. SWAIN
Master of the Christopher
Mitchell of Nantucket.
This worthy man,
after fastning to a whale,
was carried overboard by
the line, and drowned
May 19th 1844,
in the 49th Year of his age.

Be ye also ready: for in such an hour as ye
think not, the Son of man cometh.

The Bethel's ministry was and is the sea; new names are added to these plaques as the port loses its sons to the ocean. Yet this place could be a stage set, and, for all I know, John Huston's cameras might still be in the gallery, filming his 1954 version of *Moby-Dick*, while the high-ceiled chapel echoes to the plaintive hymn of Jonah's plight,

The ribs and terrors in the whale
Arched over me a dismal gloom

and Orson Welles, playing the fictional Father Mapple of Melville's story, sermonizes to his sea-bound congregation on the same biblical story,

> Yes, the world's a ship on its passage out, and not a voyage complete; and the pulpit is its prow.

Here Ishmael pays his respects to his maker, and here he listens to Father Mapple preach from a pulpit constructed to look like a ship's prow. But Huston's film – which received its world premiere in New Bedford's State Theatre, after a parade through the town led by its star, Gregory Peck – was actually made in England, and the theatrical pulpit that stands here now was commissioned in 1961 from a local shipwright to satisfy movie fans who came here expecting to see it.

Outside, the streets that Ishmael saw as dreary 'blocks of blackness' are empty of extras as I cross the road to the modern Whaling Museum, where I am greeted by the skeleton of a fifty-ton, sixty-six-foot blue whale hanging over the receptionist's desk like a gigantic children's mobile.

Washed ashore on a beach on nearby Rhode Island in 1998, this specimen was, at six years old, just a baby, but it created a giant problem. Claimed both by the museum and by the Smithsonian Institution, a compromise was reached; a leviathanic judgement of Solomon. It was agreed that the museum could have the whale, on condition that it was put on public view, visible by day and night.

In order to accomplish this feat, the whale first needed to be taken apart. The carcase was cut up into sections which were then lowered into the river in cages. For two years the minute denizens of the Acushnet ate away at the whale's flesh, until its skeleton was picked as clean as a spare rib. The reassembled result now swims through an atrium built to satisfy the Smithsonian's stipulation, an orphaned infant in a glass limbo. Incontinently, it still drips oil, like sap from a newly cut conifer or tar from a railway sleeper. The scent pervades the hall: an indefinable ocean aroma, imparting an oiliness to the air itself.

New Bedford's museum is compendious; almost every known image of the whale is represented here. Most splendid of all is Esaias van de Velde's *Whale Beached between Scheveningen and Katwijk, with elegant sightseers* of 1617, which shows just one in a series of sperm whales thrown upon the coast of the Netherlands in the sixteenth and seventeenth centuries. Such strandings were emblems of the country's fortunes at a time of flux, and in scenes of composed disaster they were replicated in engravings and even on Delft plates and tiles. They were narratives of the Dutch Golden Age – and the threats to it – and in one extravagant and remarkably accurate image, Jan Sanredam depicts a sixty-foot-long sperm whale washed up at Beverwijk on 19 December 1601.

The whale lies between land and sea; its physicality is startling, almost overwhelming. Arranged along the length of its belly are finely dressed visitors in doublets and ruffs – among them, the artist himself, seen in the foreground with his assistant holding up his cape as a screen while his master sketches. As they strike poses or perch on horseback, there is a strange, allegorical distance between them and the whale, as if they existed entirely in other dimensions. Here a whale, there the people.

Even the dogs stare.

The most prominent figure at the centre of the picture – and to whom it is dedicated – is the beplumed Prince Ernest, Count of Nassau. He was hero of the recent war against Spain, yet he uses a handkerchief to protect his aristocratic nose from the stench. Others clamber onto the whale itself; one officer plunges his sabre into its spout hole.

They crawl like ants, these humans, over and around the ravished animal. Behind its massive but now impotent tail, over which a rope has already been thrown, carriages convey more silk-clad noblemen, and tents have been set up to cater for the crowds which appear to be arriving in droves. Had it been stranded across the English Channel, this creature would have been the property of the Virgin Queen; Elizabeth I was fond of whale meat. Here in Holland, it was the subject of artists who sought to capture the strange mortality of such natural phenomena. In 1528 Albrecht Dürer, who was nearly shipwrecked, and subsequently suffered a fever which precipitated his early death when trying to reach a stranded whale 'much more than 100

Seu lætum est aliquid Patriæ, populoq Batavo,
Sive minax, veniam trepidi pacemq rogamus:
Præscia nam si quid vatum præsagia possunt;
Sunt iræ documenta tuæ, cladisq futuræ
Quot portentosum Cattorum in littore monstrum
(Tantum signa valent) clades, cædesq secutæ?
Nec minus hoc miseris dirum mortalibus omen
Vidimus obscurum Phoebi ferrugine vultum
Fœdatumq diem tenebris, Lunæq labores;
Alia laborantis, nutantia climata mundi;
Prodigia, æternis mandantur condita fastis.

fathoms long' in Zealand, reported that the local population were concerned by 'the great stink, for it is so large that they say it could not be cut into pieces and the blubber boiled down in half a year'. Such incidents seemed harbingers of death: the Scheveningen whale took four days to die, at which point its bowels exploded, fatally infecting its audience.

Full of potent signs and wonders, Sanredam's picture is framed with the apocalyptic events foretold by the coming of the leviathan. A pair of cherubs supports a cartouche containing a recent earthquake, *Terra mortus*. On either side, we see eclipses of the moon and sun, themselves flanked by halves of the severed whale, its future fate. Meanwhile Father Time looks down from one corner, and a winged Angel of Death aims his bow from the other, symbol of the plague that had recently ravaged Amsterdam. In a picture so rich in imagery, it is notable how one's attention is drawn to the animal's extended penis. Like a sixteenth-century codpiece, it makes a statement of virility, or its lack; its flaccidity is a counterpoint to the prince's upright plume, and the whale's name. From a zoologist's point of view, however, it is proof that only bull sperm whales venture this far north.

New Bedford's museum is full of whales as seen by men. Whales spouting blood as sailors ride them like jockeys. Whales belly-up, gasping as harpoons and lances are teased into their undersides. Whales painted in Hollywood style, apparently triumphant. What would Ishmael say if, while awaiting his whaling passage, he decided to loiter a little longer in the port – say, a hundred and fifty years or so – and paid his seven dollars at the cash till to cast a critical eye over this collection?

In the chapter entitled 'Of The Monstrous Pictures of Whales', our stern narrator takes issue with such 'curious imaginary portraits'. He lays the blame with the ancients as the 'primal source of all those pictorial delusions'; but the worst offender of his day was Frédéric Cuvier, brother of Baron Cuvier, the distinguished French scientist. His *Sperm Whale* of 1836 was, as Ishmael put it bluntly, 'a squash'. It was a question of attribution. Advised by the French Academy that there were no fewer than fourteen species of sperm whale, artists duly delivered images more like fashion plates of Directoire dandies, whales corseted and collared *à la mode*, sleek with fish tails, or with disproportioned bellies and misplaced eyes.

What did whales really look like? Ishmael acknowledges that there are good reasons for such glaring errors. These animals were seen in their entirety only when beached, he notes, and 'the living Leviathan has never yet fairly floated himself for his portrait . . . So there is no earthly way of finding out precisely what the whale really looks like.' The remarkable thing about his statements – which are never less than remarkable – is that they still hold true. Cetaceans remain unfathomable. The whale would stay 'unpainted to the last',

> And the only mode in which you can derive even a tolerable idea of his living contour, is by going a whaling yourself; but by doing so, you run no small risk of being eternally stove and sunk by him. Wherefore, it seems to me you had best not be too fastidious in your curiosity touching this Leviathan.

Similarly, turning the pages of old books, whaling prints resemble Renaissance masters, only with something fatally wrong: not angels announcing virgin births, or merchants' wives sitting calmly in tiled parlours, but the frenzied struggle of a gigantic animal in its death throes. The stillness of such images seems to accentuate their strangeness, to widen the gap between what they are, and what they seek to portray. In all these pictures of whales – in paint, in teeth, in wood, in sheet-iron, in stone, in mountains, in stars – never was the distance between description and actuality so great. Never have words and pictures failed us so comprehensively.

There is something about the sperm whale that leads me on, something that, even now, I find it hard to describe. No matter how many pictures I might see, I cannot quite comprehend it. No matter how many times I might try to sketch it, its shape seems to elude me. None the less, my curiosity remains, for all Ishmael's caution. And as he lingers in New Bedford's cobbled streets, calling into Carter's for some last-minute apparel before his long journey ahead – even as he readies himself for his own close encounter – my fitful and increasingly dubious guide seems to challenge me to discover why 'above all other hunted whales, his is an unwritten life'.

III

The Sperm Whale

I know him not, and never will.

The Tail, *Moby-Dick*

In some medieval past, someone pierced the head of the whale, releasing the waxy oil that filled it. As it hit the cool northern air, this hot, precious liquid became cloudy, looking for all the world like semen. Thus men came to believe that the leviathan carried its seed in its head. It may be saddled with an inelegant, even improper name, but it is also an entirely apt title, for the sperm whale is the seminal whale: the whale before all others, the emperor of whales, his imperial cetacean majesty, a whale of inherent, regal power. It fulfils our every expectation of the whale. Think of a whale, and a sperm whale swims into your head. Ask a child to draw a whale, and he will trace out a sperm whale, riding high on the sea.

But the sperm whale also bears the legacy of our sins; an animal whose life came to be written only because it was taken; a whale so wreathed in superlatives and impossibilities that if no one had ever seen it, we would hardly believe that it existed –

and even then, we might not be too sure. Only such a creature could lend Melville's book its power: after all, *Moby-Dick* could hardly have been written about a butterfly.

Scientifically, it is in a family of its own. Sperm whales – classified *Physeter macrocephalus* or 'big-headed blower' by Linnæus, the father of taxonomy, in 1758, but commonly called cachalots – are the most ancient whales, the only remaining members of the Physeteridæ which evolved twenty-three million years ago and numbered twenty genera in the Pliocene and Miocene. (In fact, Linnæus at first identified four species: *Physeter macrocephalus*, *P. catodon*, *P. microps* and *P. tursio*, but all are now known as one, with the pygmy and dwarf sperms – *Kogia breviceps* and *K. sima* – recognized as a separate family, Kogiidæ.) Relics of prehistory, they are, in one scientist's words, 'victims of geologic time . . . held in the rubbery bindings of [their] own gigantic skin'. Their nearest relation on land is the hippopotamus, although with their grey wrinkledness, small eyes and ivory teeth, they remind me more of elephants.

The sperm whale remains a class apart. Its shape itself seems somehow unformed, inchoate, as though something were missing – a pair of flippers or a fin. It is an unlikely outline for any animal, still less for the world's largest predator. To Ishmael, the whale was the ominous embodiment of 'half-formed fœtal suggestions of supernatural agencies'. Now it is seen as a 'generally benign and vulnerable creature'; from a fearful foe it has become a placid, gentle giant of the seas. The distance between these two notions is the distance between myth and reality, between legend and science, between human history and natural history. It is a mark of its magical nature – and a symbol of the fate of all cetaceans – that the sperm whale has achieved such a transformation, from wilful dæmon to fragile survivor.

Physeter macrocephalus may have been around for millennia, but we have really only known it for two hundred years; only with the advent of modern whaling, at the beginning of the eighteenth century, did man come to comprehend even an inkling of the animal. It continues to confound us. The sperm whale is a greater carnivore than any dinosaur – a fact that threatens to turn its fearsome jaws into those of an aquatic tyrannosaur - although its body is ninety-seven per cent water, just as humans are mostly made of the same liquid; we all contain oceans within us. Like other whales, the sperm whale never drinks. It has been described as a desert animal; like a camel living off its hump, its thick layer of blubber allows the whale to weather the vicissitudes of the ocean, from feast to famine. In an environment in which food stocks alter drastically, there is an advantage in being able to live for three months without having to eat, and to be able to range over huge distances in temperatures ranging from tropical to Arctic.

Truly, these are global animals. Sperm whales live in every latitude and every ocean, from the North Atlantic to the South Pacific, even in the Mediterranean. Visual surveys from planes and ships have calculated that 360,000 of them still swim the world's seas, although that is barely a quarter of the population that flourished before the age of the iron harpoon. Their love of deep water, foraging off steep continental shelves, meant that until recently only whalers – who described their quarry as travelling in *veins*, as if guided 'by some infallible instinct' ('say, rather, secret intelligence from the Deity', adds Ishmael) – saw sperm whales alive. As a result their study is still in embryo. It is as though we have hardly advanced since nineteenth-century illustrators depicted overweight whales lying on palm-fringed tropical beaches.

A SPERM WHALE.

What facts we do know cluster together like the whales themselves, defying interpretation. What colour are they? Underwater, they appear ghostly grey filtered through the ocean's blue, but in sunlight they appear brown or even sleekly black, depending on their age and sex. They may even verge on a dandified purple or lavender, with pale freckles scattered on their underbellies, leading to the pearly whiteness of the 'beautiful and chaste looking mouth! from floor to ceiling lined, or rather papered with a glistening white membrane, glossy as bridal satins'. From the side and below, this whiteness glows like a half-open fridge; an invitation, and a warning. The huge head is patchy and mottled where the tissue-thin skin is constantly peeling like old paint; it is relatively smooth, but behind, the rest of the body is furrowed and creased like a prune. This mutability gives the animal a metamorphic dimension.

From a hydrodynamic point of view, the sperm whale looks as though it were designed by an eccentric engineer. There are no concessions in its shape. Its sharp-angled flukes are not those of the sinuous and feminine humpback. It is a blunt blunderbuss of an animal; abrupt, no-nonsense. Its squareness appears to confront the water, to defy, rather than comply with the sea. Yet seen from above, its block-like head is quite narrow, wedge-shaped: this is an animal built to spend most of its life in the depths, so much so that one scientist considers it more apt to call the sperm whale a surfacer rather than a diver. Its very size allows the whale to spend long periods of time in the depths, its body being one huge oxygen tank.

Slung beneath its signature snout is the sperm whale's other most formidable feature: its lower jaw, studded with forty or more teeth which fit into its toothless upper mandible like pins in an electrical socket. These ivory canines range in size from hen's egg to massive foot-long cones too broad for me to encircle in my fingers. Sliced in half, a tooth can reveal its owner's age by counting the layers of growth like the annular rings in a tree. In the most elderly whales, the teeth are 'much worn down, but undecayed; nor filled after our artificial fashion', Ishmael observes, although, in truth, sperm whales often suffer caries. In rare instances, they also possess unerupted upper teeth, relics of ancestors who boasted a full dentition. Natural selection has left their descendants with only a lower row, as if they had misplaced their dentures during the night. That fact makes the sperm whale seem more benign; only half a monster.

The teeth are yellowy in colour; only when polished do they acquire their bright creamy whiteness, like the little ivory tusks in the carved ebony elephants my grandfather brought back from India after the First World War. Heavy in the hand, they are tactile, smooth, weighty with their benthic provenance. For all their prominence, their function is oddly obscure. One nineteenth-century writer observed that the teeth were marked with oblique scratches, 'as though made with a coarse rasp', the result, he thought, of 'corals, crushed shells, or sand' and frequent contact with the ocean floor. However, food found in the bellies of sperm whales seldom shows any tooth marks. Juveniles are eating squid and fish long before they develop teeth, and females do not produce any until late in maturity, if at all. Evidently, teeth are not necessary for sustenance. (In some cetaceans they are a positive hindrance: strap-toothed beaked whales, *Mesoplodon layardi*, have tusks which gradually grow over their jaws, creating a muzzle through which they still manage to feed.)

In his *Natural History of the Sperm Whale*, published in 1839, Thomas Beale noted that three hunted whales, one of which was blind, and the other two with deformed jaws, were in otherwise good condition, proving that not only did they not need their teeth to feed, they had no need of eyesight, either. This great predator does not chew its prey; rather, it sucks it in like a giant vacuum cleaner, as the presence of ventral pleats on its throat indicates. Some commentators have proposed that sperm whales use their jaws as giant lures, dangling them like an angler's rod and baited with bioluminescence from previous meals of squid. Beale believed that the whale hung passively in the water, waiting for its food, while squid 'actually throng around the mouth and throat', attracted as much by the 'peculiar and very strong odour of the sperm whale' as by the 'white dazzling appearance' of its jaw. However, modern science has discovered otherwise.

Addressing the conundrum of the sperm whale's head, Ishmael points out to his otherwise ignorant readers that its true shape is in no way reflected by its skull; no one who saw its bones could ever guess that the living animal possessed such a snout. To him, this is further evidence of deception on a massive scale, and in a phrenological diagnosis – all but feeling the whale's bumps – he declares that the huge forehead which lends the animal a semblance of wisdom 'is an entire delusion'. But Ishmael was himself misled, for the sperm whale boasts the biggest brain of any creature ever alive, weighing as much as nineteen pounds (as opposed to the human's seven). Quite what it does with such an organ is another matter.

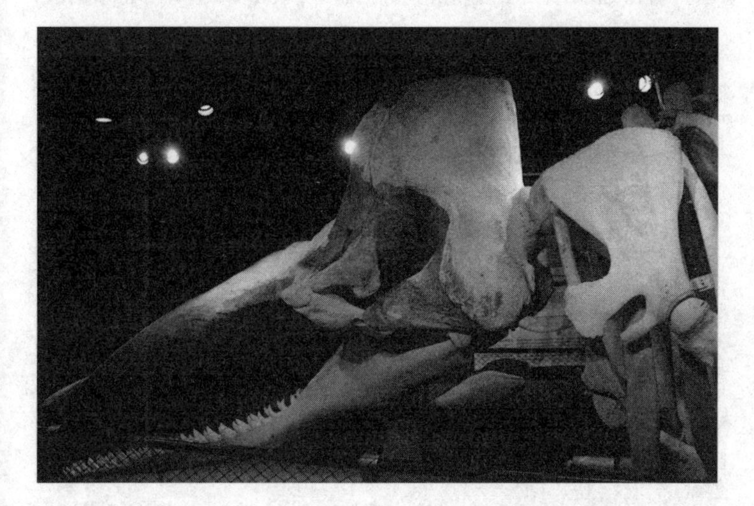

Straddling a gallery to itself in New Bedford's museum is the skeleton of a sperm whale; merely to walk around it is an intimidating experience. The skull alone is more than twenty feet long and stands higher than my shoulders. It is an essentially asymmetrical structure by virtue of its left-leaning blowhole (odontocetes possess single nostrils, whereas mysticetes have two), literally sinister (and I wonder if whales are cack-handed like me). This same quality lends an air of abstract sculpture to the complex construction of caverns and sockets created to accommodate vital vessels and nerves. One opening connects its spinal cord to its brain, another to the ears and eyes, themselves protected by the bony mass from which swings its wishbone jaw, a toothed 'portcullis', hanging 'like a ship's jib-boom'. I cannot help but agree with Ishmael: this calcium scaffolding can hardly indicate the true shape of the animal. One might tell the form of a human being from its bones, but who could imagine the reality of this creature?

As in death the enigmatic sperm whale gives few of its secrets away, so in life it sees us from another angle. Its eyes are so positioned as to prevent the animal from seeing straight ahead (although their siting on its wedge-shaped head, at the point where it narrows down to the jaw, is such that a whale can see below itself in stereoscope – presumably a useful tactic in hunting – and will swim upside-down to scrutinize objects above and, perhaps, to feed on them). For most of its life the whale must regard the world in two halves, Ishmael deduces; its head gets in the way, 'while all between must be profound darkness and nothingness to him'. It seems odd that such a powerful creature should be so benighted. This blindness is also the reason, says Ishmael, for the sperm whale's 'timidity and liability to queer frights'. A 'gallied' animal would sound deep into the ocean, beyond the reach of man and his harpoons.

In such a silent flight, the sperm whale could not be outdistanced. More than any other marine mammal, it is a master of the sea. Using its muscle-bound tail, it can power its way thousands of feet below, its paddle-shaped flippers tucked into its flanks as neatly as an aeroplane's undercarriage. And once below, it can stay down for up to two hours. To achieve this feat, a whale must spend much of its time breathing at the surface – its 'spoutings out', as the sailors called them – taking some sixty to seventy breaths in ten or eleven minutes.

> . . . the Sperm Whale only breathes about one seventh or Sunday of his time.
>
> The Fountain, *Moby-Dick*

Whereas humans inefficiently hold their breath to dive, whales supercharge their oxygen-carrying hæmoglobin blood cells before sounding, often in exactly the same spot at which they surfaced, perhaps to be sure of their survey of the food below. On these stately travels into the deep, they are accompanied by remora, sickly grey attendants suckered to their wrinkled flanks like imps; 'fish, to be sure, but not quite proper fish', they are parasites lacking individual motion, dependent on their hosts without whom they would flop to the ocean floor. Even more dæmonic are the lampreys, 'wriggling, yard-long, slimy brown creatures that repel even the zoologist'. These attach themselves to the whales with rasping mouths, leaving love-bite scars on their huge but helpless victims.

Commonly, a sperm whale will dive between three hundred and eight hundred metres, following a U-shaped trajectory. Once it has reached its chosen depth, it will swim horizontally for up to three kilometres, presumably foraging. Occasionally, the whale will dive even deeper. Dead sperm whales have been found entangled in underwater cables 1,134 metres down – although that figure does not measure the drowning agony of the whale, its jaw caught in the insulated wire.

In 1884, a cable-repairing steamship operating off South America pulled up a cable in which a dying whale was trapped, its entrails spilling out; the wire itself was found to be bitten in six places. In another insight gained at mortal expense, a sperm whale caught south of Durban in South Africa in 1969 was found with the remains of two *Scymodon* sharks in its belly. Since such fish are bottom dwellers feeding at three thousand metres, this was proof of the whales' extraordinary diving abilities. Much of what we know about sperm whales was discovered by those whose primary interest was to kill them. Whales died that men might describe them.

Nor are they easily replenished. The sperm whale has the lowest reproductive rate of any mammal – females produce single calves only once every four to six years. It is also the most sexually dimorphic cetacean: males may be twice the size of females. The sexes live apart for most of their lives, the males growing larger and all the more attractive to their potential, if fleeting, partners. This also has the benefit of maintaining the supremacy of their species: the vast distances they travel ensure that the global population of sperm whales is surprisingly genetically similar.

Moving south to breed, males fight for the females' favours. The distorted jaws – some even tied in knots; Moby Dick's own mandible is described as sickle-shaped as it scythes off Ahab's leg – that Beale saw are evidence of these ferocious but short-lived battles, as are teeth marks on the animals' heads, backs and bellies. Although they have no territory to defend like rutting stags, whales will take bites out of each other's blubber, ramming one another with pugnacious foreheads which in males become almost obscenely extended.

Successful suitors mate belly to belly, with females underneath – *more hominum*, in Ishmael's discreet words. Gestation lasts fifteen months; calves are nursed for at least two years, sometimes communally, and thirteen-year-olds have been known still to suckle. 'The milk is very sweet and rich,' says Ishmael; 'it has been tasted by man; it might do well with strawberries.' Lacking lips, calves take the milk into the side of their mouths as it is squirted from their mothers' teats, a technique first identified by the surgeon Sir William Wilde, father of Oscar.

Sperm whales have the most complex social structure of any animal other than man. Like other toothed whales, they travel together, separated by sexual maturity into reproductive and bachelor groups. Females and immature whales swim together in groups of twenty to thirty, dispersed over a wide area; they prefer warmer waters, possibly because fewer killer whales – their only natural predators – are found at such latitudes. Communal care confirms these extended bonds: when a mother dives for food, she will leave her calf – which cannot yet follow her – in the care of other females or juvenile males in a cetacean crèche. Large males have been seen gently carrying calves in their mouths, although the fact that they simultaneously exhibit extended penises probably means that this has more to do with mating than nursing.

In their teens and twenties, young males join bachelor groups, as though entering a rite of passage. They attain maturity at nineteen (females become sexually mature as early as seven years old), but do not mate until their twenties. They travel further in search of prey; adult males roam more than forty degrees latitude north or south, forming loose concentrations spread over two hundred miles or more. Eventually these groups reduce in size until, in middle age, the males become solitary, roving as far as sub-polar seas to find new feeding grounds before returning to warmer waters to mate.

In the interests of order, the whalers subdivided their subjects, applying human terms of trade and organization to marine mammals:

> Pods, or gams: up to twenty whales
> Schools, or shoals: twenty to fifty whales
> Herds, or bodies: fifty whales, or more

Single bulls were schoolmasters, groups of females were harems, and young males, bachelor schools of 'forty-barrel bulls'. Ishmael gives us a memorable description of a nursery into which the *Pequod* sails. He looks down through limpid waters, to where

> another and still stranger world met our eyes as we gazed over the side. For, suspended in those watery vaults, floated the forms of the nursing mothers of the whales . . . and as human infants while suckling will calmly and fixedly gaze away from the breast . . . even so did the young of these whales seem looking up towards us, but not at us, as if we were but a bit of Gulf-weed in their new-born sight.

None of this, however, prevents the crew from laying into the innocent scene. It is one of the cruellest aspects of its historical fate that this most hunted of whales is built for a long life, a longevity indicated by the slow beating of its huge heart at ten times a minute; a shrew, whose heart beats one thousand times in a minute, lives for just a year. It is as if the animal's life history had been slowed down by virtue of the millions of years its species has existed. At forty-five, a sperm whale is middle-aged, and has achieved its optimum size; like a human, it enters old age in its seventies. Females live into their eighties and perhaps to one hundred years or more, although none is known to have given birth after their forties. Rather, these matriarchs assist other females 'in ways we do not yet understand', as Hal

Whitehead, one of the great modern experts on sperm whales, says. He calls these older females 'sages', raising images of elderly, grey-haired grandmothers, teaching their sons and daughters how to raise their children and passing on memories of good feeding grounds.

Given their slow breeding and the centuries of hunting they have endured, it is a testament to their evolutionary success that the *Physeter* should remain so ubiquitous throughout the world's oceans; among mammals, only killer whales and humans achieve such a cosmopolitan reach. Although they adhere to deep water, sperm whales have been seen off Long Island, almost within the city limits of New York, while others swim not far from the coast of Cornwall or Norway. These are generally lone bulls, but other whales may travel in schools of hundreds or even more, numbers 'beyond all reasonable conception' to Frederick Bennett. Whalers would suddenly come across huge herds of these enormous animals, like buffalo on the plain. Dr Whitehead too compares them to elephants, roaming the ocean's savannah, with similar social structures and mutual dependencies – even the same highly modified and very useful noses.

And as they roam the oceans, whales observe neither night nor day. Like all whales, they are voluntary breathers, and must keep half their brains awake while they sleep, during which – if dogs are anything to go by – they certainly dream. Sometimes they hang perpendicularly like bats, blowholes to the surface, dozing in a drowsy cluster after feeding. Sperm whales exhibit social skills that go far beyond the herding instinct. They enjoy the contact of their bodies, spending hours slowly rolling around one another just below the surface. 'They seem to love to touch each other,' says Jonathan Gordon of this underwater ballet. 'It is not unusual to see animals gently clasping jaws.'

Such cohesion extends to self-defence. Forever on the move, whales will swim in ranks, 'like soldiers on parade', seeking safety in numbers, diving in clusters to feed, synchronizing their soundings as security against predators. Even such fiercely armed animals are vulnerable to attack from orca – more especially so in the three-dimensional hunting ground of the ocean where there is no place to hide, and where a victim can be approached from any angle. Here their only refuge is each other.

Threatened sperm whales will stop feeding, swim to the surface, and gather to each other in a cluster. Assembled nose to nose around their calves, they form a tactical circle known as a 'marguerite', bodies radiating outwards like the petals of a flower. Thus they present their powerful flukes to any interlopers, protecting their young in a cetacean laager. In an alternate version, they touch flukes, heads out and jaws at the ready. Besieged whales will maintain these positions silently, unmoving. If a whale is separated from the circle, one or two of its companions will leave its safety to escort the animal back to the formation, risking their own lives as the killers take great chunks out of the sperm whales' flesh, foraging like packs of wolves. These are, writes one naturalist, '"heroic" acts by almost any definition'.

Ironically, while such techniques are successful in repelling orca, they also made the animals more susceptible to slaughter by man. 'The females are very remarkable for attachment to their young,' Beale observes, 'which they may be frequently seen urging and assisting to escape from danger with the most unceasing care and fondness.' If one were attacked, 'her faithfull companions will remain around her to the last moment, or until they are wounded themselves'. This was known as 'heaving-to' by whalers, who capitalized on their prey's fatal tendency to foregather when endangered, and destroyed entire schools 'by dextrous management'. 'They did not swim away or dive,' wrote an observer of a twentieth-century hunt. 'The gunner, therefore, took the whales very easily, starting with the largest one.' As Beale adds, poignantly, 'The attachment appears to be reciprocal on the part of the young whales, which have been seen about the ship for hours after their parents have been killed.'

To humanize the whale oversteps boundaries; but when entire families follow a stricken relative to strand on a beach, or when a wounded female, mortally gashed by a ship's propeller, is borne up by the shoulders of her fellow whales, it is difficult to resist the pang of emotion. They are truly gentle giants: as elephants are supposed to bolt at the sight of a mouse, so sperm whales can be faced down by a pod of militant dolphin. The appearance of a seal, or even the click of a camera, may send them scurrying. It is almost as though, as Dr Whitehead remarks, the whale sees its own habitat as a dangerous, even frightening place.

Yet these are carnivorous animals, voracious in their appetites. They eat mostly cephalopods, but also take tuna and barracuda; entire thirty-foot sharks have been found in their bellies. And they consume in enormous proportions, taking from three to seven hundred squid a day: worldwide, sperm whales eat one hundred million metric tons of fish a year – as much as the annual catch of the entire human marine fishery.

Diving deeper than any other mammal, we simply do not know how sperm whales behave in the ocean's depths. We know what they eat, because we find it in their stomachs; but we don't know how it gets there. Sound is certainly important to their sustenance. Although they lack a voice box – as Thomas Beale noted, 'The sperm whale is one of the most noiseless of marine animals . . . it is well known among the most experienced whalers, that they never produce any nasal or vocal sounds whatever, except a trifling hissing at the time of the expiation of the spout' – the whale possesses the largest sound system of any animal, using one-third of its body to create the loud clicks that it constantly emits when hunting. The whale's oversized nose is in fact a huge and highly efficient squid-finder.

As bats send out sonar to find flying insects, so sperm whales send out similar, if rather louder, pulses to locate their prey. Their characteristic clicks are produced by the expansion and contraction of 'blisters' on their nasal sacs. It is a remarkably complicated sequence, as Dr Whitehead explains. Two nasal passages run from the external blowhole, the left and the right. The left runs directly to the lungs, but the right passes through a distal air sac via a kind of valve known as the *museau du singe*, or 'monkey's muzzle'.

Sound is initially generated by air being forced through this valve – not unlike the clicks you can make by hitting the roof of your mouth with your tongue – then passes through the animal's upper spermaceti organ or 'case' before bouncing off another, frontal air sac set at the back of the skull – a bony sound mirror, in effect. This is then redirected and broadcast through a series of acoustic lenses in the 'junk', the lower oil-containing organ in the whale's head. Thus the strange mechanism of the sperm whale's nose acts as a living amplifier. Some sound also bounces back and forth along the case, producing a second pulse. As this inter-pulse interval is equal to the length of the case, the actual sound created by the whale – the pulses between its clicks – may be a measurement of its physical size; one may tell the length of the animal from the inter-pulse interval, just as the bigger the whale and its head, the more powerful its clicks. Breeding males may size each other up from their clicks, and can tell each other's sex by the same sound; they are as much a tribal definer as the click speech of the Xhosa of South Africa.

The clicks, which can be heard for many miles, are important for navigation and communication. They extend the whale's sensory map far beyond its own body, and their speed and variation change from group to group, as an English dialect changes from Yorkshire to Hampshire. This allows individual whales to identify and communicate with mem-bers of their family, evenas they use the earth's magnetic fields to map out their subaquatic terrain, the peaks and valleys of the oceanic abyss in which they are effortlessly at home. And as they dive – often in an informal group – they use their clicks to locate and scan, with extraordinary

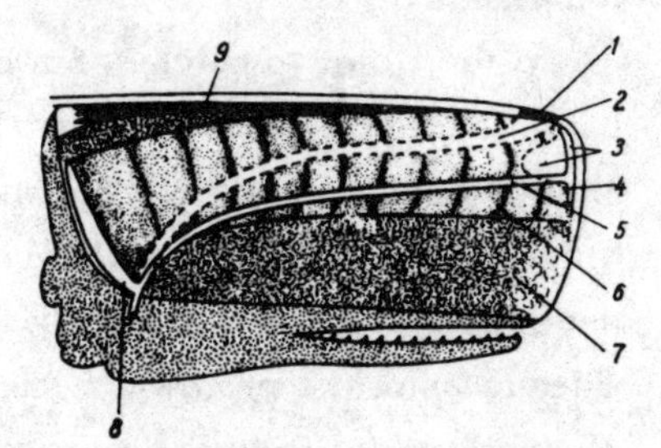

FIGURE 65. Structure of spermaceti organ and system of epicranial nasal passages of the sperm whale:

1 — blowhole; 2 — left nasal passage; 3 — distal cavity; 4 — pouchlike evagination of distal cavity; 5 — right nasal passage; 6 — upper spermaceti sac; 7 — lower spermaceti sac; 8 — frontal cavity; 9 — muscular-tendinous layer.

precision, the distance, presence and nature of their prey. It is thought that a whale can 'see' into its prey, diagnosing it – even to the extent that it can tell if it is pregnant. The returning clicks are 'heard' through the dense, hard jaw bone – the same bone from which Ahab's false leg is carved – and which acts as a listening device in its own right, conducting sound through bioacoustical oils directly to its eardrums. The whale's external ear is largely useless; the animal hears through its body itself.

The deeper it dives, the more effective the whale's senses are, away from the chatter and interference of the world above. A sperm whale can create a two-hundred-decibel boom able to travel one hundred miles along the 'sofar' channel, a layer of deep water that readily conducts noise. It seems strange that such a physically enormous creature should rely on something so intangible; but bull sperm whales, by virtue of their larger heads, generate sounds so powerful that they may stun or even kill their prey. These directional acoustic bursts, focused through their foreheads and likened to gunshots, are the equivalent, as one writer notes, of the whale killing its quarry by shouting very loudly at it.

In their own researches, Soviet scientists, whose nation's enthusiastic hunting of the sperm whale in the twentieth century allowed ample opportunity for such study, suggested that in order to hunt in the depths where only one per cent of sunlight penetrates below two hundred metres, the whale uses a 'unique video-receptor system . . . which lets the animal obtain the image of objects in the acoustic flow of reflected energy even in complete darkness'. In other words, the sperm whale can see its prey in sound. And just when you think nothing else about this animal could confound you, another theory proposes that the whale's sonic bursts, and the movement of its head, may cause plankton in the deep water to emit their bioluminescence. In the utter darkness, the leviathan may light its own way to its lunch.

Even as you leave the Tube station, you remain an underground passenger, conducted through a tiled tunnel before emerging into the shadow of an extravagant cathedral of science. Clinging to the terracotta façade – itself layered to resemble geological strata – is an industrial bestiary: heraldic griffins and scaly medieval fish and, most frightening of all, grinning, toothy pterodactyls, with their obscene storks' beaks and glaring gargoyle eyes and their leathery wings wrapped about them.

In the gothic nave, children mill about a blackened diplodocus nonchalantly waving its whiplash tail. A hundred years ago, they would have been greeted by another monster, for here stood the skeleton of a sperm whale, guarded by what appeared to be a Victorian policeman as if it were a prisoner at Pentonville.

Skeleton of Sperm Whale.

The route comes back like a lost memory. I walk past ichthyosaurs sailing through long-vanished Triassic seas and moth-eaten fauna of the savannah and the jungle, displays out of a dead zoo. Abruptly, the corridor turns into a space more like an aircraft hangar than a museum. There, hanging like one of the model aeroplanes I used to suspend from my bedroom ceiling, is the blue whale, the largest object in the Natural History Museum.

Contrary to the usual tricks of childhood recollection, it is actually bigger than I remember. Nearly one hundred feet long from the tip of its nose to its twenty-foot flukes, the whale could easily accommodate a large household within its interior. There is something fairy-tale about it, an invention of the Brothers Grimm: its huge mouth has a faint grin, and its disproportionately small eye stares out from its wrinkled socket, part amused, part pleading. Even Linnæus's name for it is a little Swedish joke: *Balænoptera musculus – Balæna* meaning whale, *pteron*, wing or fin, and *musculus*, both muscular, and mouse.

I was fooled by this model then, as visitors are now, for this wood and plaster reconstruction is only an approximation of a blue whale, a much more streamlined animal than this bloated model gives credit. Constructed in the 1930s, before anyone had seen an entire living whale in its element, the creators of this whalish effigy relied on carcases hauled out of the water, where they lay deflated like old inner tubes, incapable of bespeaking their true beauty. Like the dinosaurs of Crystal Palace – where we went on another family pilgrimage, to see concrete iguanodons and plesiosaurs stranded in a suburban park – London's magnificent whale is an object of error and mystification. As a boy, I assumed that inside the model was the animal's skeleton, like a cathedral tomb containing the bones of a saint. In fact, the whale is hollow, and was made using plaster and chicken wire over a wooden frame constructed on site – as if the great hall had been built around it.

The idea of a new Whale Hall for the museum had been posited as far back as 1914, but war put a stop to it. The project was revived in 1923, when the museum's pioneering director, Sidney Harmer, called the Trustees' attention to 'the inadequacy of the exhibited series of the larger whales. The subject of whaling is very much in the air at the present time,' he noted, and he reminded the Trustees that they had 'frequently expressed their sympathy with efforts to protect whales from extinction'.

Warming to his theme – on three sheets of pale blue foolscap paper – Harmer declared that 'under such circumstances it would be natural to expect that such species as the Greenland Whale, the Blue Whale and the Humpback Whale would be illustrated in the Whale Room . . . to give the visitor a satisfactory idea of what these three important species are like'. It was even suggested that government grants for the relief of unemployment and men disabled by the war might be used. However, the primary reason for the new hall was to promote the work being done by the *Discovery* expeditions in South Georgia, where scientists were conducting their investigations alongside the British whaling fleet in the Southern Ocean.

It took almost a decade for Harmer's spectacular vision to be realized. In June 1929 the new hall was announced – complete with a glass roof framed with steel girders, in the Modernist style – but was not completed until 1931. To fill this grand new space, a life-size whale was proposed, and so in 1933 the museum decided to commission a Norwegian engineer to procure a blue whale, hang it by its tail in an engraving dock, and take a mould of it. The expense of this ambitious scheme was to be allayed by selling the blubber and by marketing models made from the mould to American museums; however, its decidedly 'experimental nature' meant that it too was abandoned.

Five years later, in April 1937, the museum's Technical Assistant and taxidermist, Percy Stammwitz, suggested that he should make the model in the hall itself. Stammwitz and his son, Stuart, spent nearly two years creating the blue whale, to measurements taken by the scientists in South Georgia. Giant paper patterns, like a dressmaker's kit, were used to cut out transverse sections in wood, which were then connected at three-foot intervals with slats. Over this armature wire netting was laid to take the final plaster coat; Stuart himself would paint the whale's eye. It was a long and laborious task, and during its construction the workers used its interior as their canteen – much as Benjamin Waterhouse Hawkins had given a New Year's Eve dinner in his half-built iguanodon in 1853, a party of scientists that one periodical portrayed as modern Jonahs swallowed by the monster.

As it rose from its timber foundations, the model resembled a huge ship whose keel had been laid down in the museum's hall, an ark ready for the launch to save the museum's species before the flood; or perhaps an inter-war airship, about to be inflated with helium for a transatlantic crossing. Indeed, when it was suspended from the ceiling, painters working on the whale complained that it swayed so much that it made them seasick.

The finished article looked so realistic that *The Times* thought it would 'no doubt be mistaken for a "stuffed" whale by the casual visitor'. On its completion in December 1938, just before the outbreak of war, a telephone directory and coins of the realm were placed inside the model as a kind of time capsule. Thus, on the eve of host-ilities, the placid whale became a memorial to a brief peace, a cetacean cenotaph. It was a giant

good luck charm, too, for the warders who put pennies on its flukes to encourage visitors to do the same, much as they might throw coins in a fountain for luck. When the museum closed for the evening, the warders scooped up the takings and spent them in the pub.

Now there is a notice on one side,

> Please do not throw coins on the whale's tail.
> It causes damage. Thank you.

and, next to it, a twenty-pence and a ten-pence piece lie on the plaster flukes.

Other models were made to supplement the display, Stammwitz's initial attempts to stuff dolphins having proved as unsuccessful as earlier attempts to mould a blue whale. They have been replaced in turn by a flotilla of fibreglass cetaceans, from a tiny Ganges River dolphin to a primeval-looking Sowerby's beaked whale, all following their leader as if one night she might break open the wall of the gallery and guide her charges down to the Thames and out to sea. Until then, there they hang, biding their time, watching the school parties with their beady glass eyes.

Below the Whale Hall, in the belly of the building, Richard Sabin, the curator of sea mammals, takes me through automatic doors that lock like a spaceship behind us, sealing the climate-controlled area from the world outside. I follow him, past ranks of giant grey lockers reaching from floor to ceiling. As he opens door after door, their contents are revealed: sections of cetaceans preserved in alcohol and labelled with their Latin binomials, *Phocœna phocœna, Tursiops truncatus, Balænoptera physalus.* One container, the size of a small fish tank, holds a humpback fœtus; with its mouth agape and its pallid skin, it looks more like a rubber toy.

The end of the corridor opens into a wide room lined with shelves on which stand jars of pale brown liquid, a sharp contrast to the flickering white hum of the lights overhead. Crammed into each glass column is an animal, ghoulishly bottled like a pickled gherkin. A spiny anteater's spikes twist as it tries to climb out of its transparent prison with its rodent paws. A severed shark's head sits at the bottom of a wide jar, staring reproachfully. Plunged in another is the scaly carcase of a coelacanth, still swimming in seas tinted tobacco by the immensity of time.

It is the stuff of my nightmares, and as I reach the end of a row of specimens – some collected by Darwin himself, and all ordered and classified with handwritten luggage labels as if ready for transit elsewhere – I back away from a big, bug-eyed bony fish which someone has left lying nonchalantly on the side, only to find my way blocked by a series of closed metal vats like pans in a canteen kitchen, all the more intimidating for the photocopied labels that indicate their invisible contents: entire dolphins and infant whales. None of these terrors, however, can compare to the gigantic plate-glass tank that runs half the length of the room, supported on bier-like struts. Inside, suspended in a mixture of formalin and sea water, is *Architeuthis dux* – the giant squid, mythical enemy of the sperm whale.

It looks strangely spectral as it lies there, the faintly green glow a pale mockery of its ruddiness in life. Rudely yanked to the surface by Falklands fishermen in the Southern Ocean, it was frozen like a giant fish finger and shipped to Hull before being brought here, to the cellars of South Kensington. At twenty-eight feet long, this specimen is by no means the largest: in 1880, a squid measuring sixty-one feet was caught in Island Bay, New Zealand. Some may grow even larger. Nelson Cole Haley, sailing on the whale-ship *Charles W. Morgan* from 1849 to 1853, claimed to have seen three huge squid swimming together off the north-west coast of New Zealand, one of which he estimated to be three hundred feet long.

'One might say this is a big fish story,' acknowledged Haley of this monstrous procession; but he had seen many whales and other creatures, and 'although I might have been frightened at what I saw, I had not lost my head so much but I could use my poor judgement about their appearance as well as ever'. He had no doubt that what he saw were 'wonderful monsters of the deep'. Science may yet confirm Haley's apparitions: recent acoustic studies have identified a 'bloop' sound from the depths which could only be made by a very large animal, and which may be a massive squid hundreds of feet in length, far bigger than a blue whale.

To sailors, these creatures were the original kraken, the sea monsters of myth, 'strange spectres' believed able to drag entire ships down to the deep. It was as though nature had created a fitting opponent for the whale. On her own hunt for Moby Dick, the *Pequod* encounters a 'great white mass' rising lazily to the surface, a creature so large that it becomes a living landscape: 'A vast pulpy mass, furlongs in length and breadth, of a glancing cream-colour, lay floating on the water, innumerable long arms radiating from its centre, curling and twisting like a nest of anacondas, as if blindly to clutch at any hapless object within reach.'

As the word 'whale' evokes poetic wholeness, so 'squid' seems expressive of fragmentary, faceless evil; and as this 'unearthly, formless, chance-like apparition of life' sinks with a 'low sucking sound', Ishmael seems to shudder, too. 'So rarely is it beheld, that though one and all . . . declare it to be the largest animated thing in the ocean, yet very few of them have any but the most vague ideas concerning its true nature and form; notwithstanding, they believe it to furnish to the sperm whale its only food.' But here in a London basement, the monster lies embalmed in its glass coffin, a legend reduced to the status of a dead fish.

It is an enormous intestinal tangle of flesh, frayed by its harsh treatment in the trawl. From its long mantle eight arms reach out in a now mushy cordage; they are studded with vicious circular suckers and barbs that could brand a whale's hide. Nestling at their roots are the squid's mandibles, hard and strong and shiny as a parrot's beak and made of chitinous material; as phallic as it is, there is more than a little of the *vagina dentata* about this monster. In its removal from the dark oceanic columns to this controlled vitrine, its huge eyes, more than a foot in diameter to allow in optimal light, have shrunk into their sockets, depriving the specimen of whatever character it once possessed, blinding it to its fate. Cephalopods have highly developed nervous systems; one reason for the animal's beak is the need to chew up its food into smaller chunks; as the œsophagus passes perilously close to the brain, an ill-considered meal might damage it. These are truly alien animals: squid also possess two hearts.

Feeling their way ahead, a pair of twenty-foot tentacles extend beyond the body, at least as long as the animal again. Far from being a passive victim, Soviet scientists suggested that the giant squid may actively wrap its tentacles around a sperm whale's head, clamping shut its jaws and even attempting to seal its blowhole, the dread of every cetacean. Few humans can claim to have witnessed such a battle. In his book, *The Cruise of the Cachalot*, Frank Bullen tells how the New Bedford whaleship on which he was serving was sailing in the Indian Ocean. Late into the night watch and under a bright moon, he saw a great commotion in the sea, far off. At first he thought it might be an erupting island. Then, through field glasses, he saw a great sperm whale battling a giant squid. The cephalopod's arms had created a kind of net around the whale's black columnar head, while the whale was mechanically chewing its way through its assailant. Bullen woke the captain to come and see this once-in-a-lifetime sight; his master merely cursed him and went back to sleep.

Such scenes may be the stuff of horror movies; but in that as yet unphotographed or filmed contest of snapping beaks and tearing teeth – a voracious and infernal coupling – the gelatinous accumulation of ganglions and sinews can be scooped up in the whale's own monstrous jaw and, as the animal's arms writhe to evade its fate, it is swallowed alive. (The squid's classical defence, the ink cloud, is useless in the face of a predator that can 'see' in the dark; although the pygmy sperm whale – a compact version of its cousin – excretes a thick reddish-brown liquid from its guts when startled, as if to emulate the method employed by the prey on which it dines.)

In a nearby jar lie lumps of squid flesh and beaks retrieved from the belly of a sperm whale by the *Discovery* expedition, as the label floating inside, inscribed in sepia, notes. In this underground laboratory, battling foes have been bottled for posterity. Hunted whales have even been found with squid still alive in their stomachs; confirmation of the existence of *Architeuthis* came when dying whales vomited up pieces of their arms and tentacles. Nor are they a rare meal for *Physeter*: ten per cent of the diet of sperm whales off the Azores consists of giant squid, and in the Antarctic, colossal squid – *Mesonychoteuthis hamiltoni*, with eyes as big as basketballs – are eaten by sperm whales, their only predators. The extraordinary nature of its quarry only underlines the abiding mystery of the hunter, feeding day and night, forever stoking the insatiable furnace of its metabolism.

Upstairs, corridors which only an hour before were filled with chattering school children have fallen quiet. I can hear the distant hum of a vacuum cleaner as I make my way past galleries of long-dead animals, past the blue whale and the dark skeletons hanging above it. Now, in the silence, they seem harmless and foreboding at the same time, resonant with what they once were. I leave by the main doors – only to find my way barred by the museum's locked gates.

I have visions of spending the night inside the museum, with the dinosaurs and stuffed tigers with their yellowing teeth and glass eyes. I think of the corner of the grounds where, until just before the war, it rendered its own specimens in pits of silver sand. Here carcasses were prepared for articulation and display, lowered into the sand where rain would percolate through, speeding up a process of decay that might take two years or more. Photographs show sperm whales being hauled out of a kind of animal dry dock, although they look to me like bodies being pulled from blitzed buildings. Only when local residents complained about the smell was the practice put to an end. It is hard to believe – as I eventually find my way into the bright lights of Knightsbridge – that behind the gothic façade dead whales once lay, tended by a boiler-suited scientist, who looked more like a gardener engaged in double-digging a trench – only with a cigarette purposefully in his mouth, presumably to counter the stench of the rotting animal at his feet.

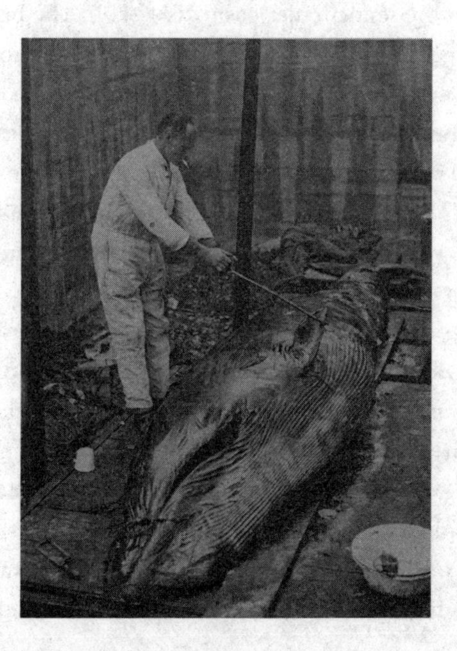

Other whales are inhabitants of the superficial waters, connected to the sun and the waves. The sperm whale is a denizen of the deep, spending half its life feeding on blind-eyed creatures of the abyss. Yet as dark as it is, *Physeter* once provided the essential element of light. For two centuries or more, that same hooded head provided luminescence for drawing rooms and street lamps from Kensington to Kentucky. The very unit of light, the lumen, was measured from a pure white spermaceti candle, one candlepower being equivalent to the burning of one hundred and twenty grains of wax per hour. As it did not freeze, sperm oil could be used in lamps during the winter, as well as a lubricant for watches and other fine instruments. The whale was itself a manufactory, of strange substances and of human fortunes.

The sperm whale's head contains two reservoirs of fluid which sit in the semicircular basin of its skull. The uppermost is the spermaceti organ, or *case*, a long, barrel- or cone-shaped structure surrounded by a muscular sheath and containing a spongy network of tissues saturated with oil. This lies on top of the second chamber, the *junk*, divided from it by the right nasal passage, also filled with oil. It is this precious semi-liquid that defines the whale – still more its historical value – and yet which its terse-lipped owner declines to explain.

In the absence of any such elaboration, science steps in. Or at least, it tries to. One theory is that the whale's head is an enormous buoyancy aid. The oil's density and viscosity changes with temperature; thus, as the whale draws in cold water along its right nasal passage, it cools the oil which becomes heavier in the process (unlike water, which gets lighter as it freezes). Warming the organ with its body heat, the effect is that of wax in a lava lamp, allowing the whale to rise and fall at will. But this elegant hypothesis is disputed, and others consider that the oil's primary function is to act as a focus for the whale's powerful sound system. In effect, its ability to carry sound turns the animal's head into a highly directional loudspeaker, allowing it to broadcast its presence.

Ishmael assigns a more sensual role to this magical liquid. In one of *Moby-Dick*'s most extraordinary chapters, 'A Squeeze of the Hand', he and his shipmates sit round a tub of spermaceti, squeezing lumps out of the cooling oil.

> Squeeze! squeeze! squeeze! all the morning long; I squeezed that sperm till I myself almost melted into it . . . and I found myself unwittingly squeezing my co-laborers' hands in it, mistaking their hands for the gentle globules. Such an abounding, affectionate, friendly, loving feeling did this avocation beget; that at last I was continually squeezing their hands, and looking up into their eyes sentimentally; as much as to say . . . Come; let us squeeze hands all round; nay, let us all squeeze ourselves into each other; let us squeeze ourselves universally into the very milk and sperm of kindness. Would that I could keep squeezing that sperm for ever!

Tellingly, this is followed by the even odder account of 'The Cassock', in which Ishmael describes a 'very strange, enigmatical object . . . that unaccountable cone . . . nigh a foot in diameter at the base, and jet-black as Yojo, the ebony idol of Queequeg'. Only the assiduous reader would realize that he is talking about the whale's penis. In a bizarre ritual, the 'mincer' removes the giant foreskin, 'as an African hunter would the pelt of a boa', and turning it inside out, stretches it and hangs it up to dry. He then cuts two armholes in the 'dark pelt' and puts it on. 'The mincer now stands before you invested in the full canonicals of his calling,' says Ishmael, 'arrayed in decent black . . . what a candidate for an archbishoprick, what a lad for a Pope were this mincer!' (Harold Beaver, a later editor of *Moby-Dick*, goes so far as to say that 'this peculiar "mincer" . . . proves to be a mincing queer' and 'this "cassock", turned inside out, spells "ass/cock" in the rigging'.)

Whether or not such a rite ever happened on board a whaleship – and it may well be a figment of the author's mischievous imagination – it is 'the most amazing chapter in an amazing book', wrote Howard P. Vincent, although he could not bear, in 1949, to discuss it further, beyond noting that 'ninety per cent of Melville's readers miss entirely the meaning of "The Cassock"'. Other writers were less coy about the sexual symbolism of the whale. D.H. Lawrence had already dubbed the sperm whale 'the last phallic being', and in 1938 W.H. Auden wrote of Ahab and 'the rare ambiguous monster that had maimed his sex' – a reference to an incident in which the captain was found one night sprawled and insensible on the ground, 'his ivory limb having been so violently displaced, that it had stake-wise smitten, and all but pierced his groin'. It was as if, in this entirely masculine world, men must sexualize the whale to make it submit – just as they might be subsumed by it in turn. By the 1970s Harold Beaver was declaring the same animal 'both bridal chamber and battering ram . . . a true amphibium, dual-sexed as Gabriel's "Shaker God incarnated"'. The protean whale had become a phallus itself, but also a spermatozoid, gigantic and seminal at the same time.

Given such mysterious and symbolic attributes, such legendary enemies and such iconic status, it is little wonder that the sperm whale was a fated beast, condemned to be the quarry of man. The blue whale and the finback were too fast, the humpback unproductive. It was the sperm whale – immediately recognizable by its angled spout, by its predilection for lying at the surface and, most paradoxical of all, by its essentially shy nature – that offered itself up as a sacrifice for all other whales: a silent, honourable champion.

FISH
FRESH FISH
ICE

IV

A Filthy Enactment

Who aint a slave? Tell me that.

Loomings, *Moby-Dick*

Housed in its own vaulted, purpose-built hall is New Bedford's grandest exhibit: a half-scale replica of a whale-ship. Even allowing for its reduced size, the confined lower decks of this vessel are intimidating. They resemble nothing so much as the slave ships of the age: the one designed to carry the harvest of dead whales; the other to convey living souls. In a nearby cabinet is a much smaller specimen: a framed daguerreotype of a handsome man with a sweep of sleek wavy hair, fine cheekbones and serious, querying eyes; he wears a dandy's high-collared shirt and tie and an elegant dark coat. But this composed figure was the fomenter behind the campaign to abolish slavery – in a city that shackled men to the pursuit of the whale.

In 1838 Frederick Douglass, the son of an enslaved mother and a white father whose name he never knew, escaped from Baltimore dressed as a sailor. He arrived in New Bedford where, for four years, he lived and worked, rolling casks, stowing ships,

sawing wood, sweeping chimneys and labouring at a blacksmith's bellows till his hands were like horn. Ishmael claimed that 'a whale-ship was my Yale College and my Harvard'; for Douglass and his brethren, 'the ship-yard . . . was our school-house'.

Like the rest of America, New Bedford is a place made up of other places. If more white Americans were descended from pickpockets and prostitutes than from the Pilgrim Fathers, then, as Ishmael informs us, 'not one in two of the many thousand men before the mast employed in the American whale fishery, are Americans born'. While America's railroads were built by Irish navvies, its dirty business of whaling was done by Africans and Indians or Azoreans and Cape Verdeans. The heroes of the harpoon were more likely to be men of colour than sons of the *Mayflower*.

By the second quarter of the nineteenth century, one in twenty New Bedfordians was black, a greater proportion than that of New York, Boston or Philadelphia. 'In New Bedford,' marvels Ishmael, 'actual cannibals stand chatting at street corners; savages outright; many of whom yet carry on their bones unholy flesh. It makes a stranger stare.' The South End of town was known as Little Faial for its Azoreans; another downtown neighbourhood was named New Guinea after its inhabitants. On these shingled and clapboarded New England streets a dozen languages could be heard and dark figures seen, fellow countrymen of Queequeg, Tashtego and Daggoo, the Polynesian, American Indian and African-American harpooneers of the *Pequod*. Visiting in 1917, Mary Heaton Vorse saw an 'illusion of the South' about the port, with its 'Bravas' or Cape Verdeans and entire neighbourhoods in which white people were the foreigners; where children stared back, and 'a splendid Negress with thin Arab features . . . checked her stride to wonder about us'.

Black sailors were engaged by owners who did not ask questions, or whose Quaker beliefs opposed slavery. Some rose to become captains or mates. Others succeeded in supply industries: Lewis Temple of New Bedford invented the toggle-iron harpoon, with its ingeniously hinged head. But below deck, bunks were still segregated and conditions were such that by the end of the century only men of colour could be persuaded to sign up; hence the preponderance of black faces in photographs of whaling crews. Charles Chace, one of New Bedford's last whaling captains, kept two loaded pistols in his cabin in case of trouble – so his descendant told me – and when his Cape Verdeans were discharged with a suit of clothes and a ten-dollar bill, many gave up their African names and, like slaves, adopted their master's, for the sake of conformity with their new home.

New Bedford owed at least part of its success to its communications with the rest of America; the same year that Frederick Douglass arrived, the city was connected by rail to the New England network. But for Douglass and for Henry 'Box' Brown – who was smuggled out of the South in a crate, emerging at the other end as a human jack-in-the-box – New Bedford was a vital stop on another network: the Underground Railroad, an invisible system secretly helping thousands of slaves to escape to the North and Canada. A port was the perfect place for such illicit trade; and whaling offered a tradition of disguise as well as employment. For Douglass and his fellow fugitives, New Bedford's transience itself was a kind of liberty: 'No coloured man is really free in a slaveholding state . . . but here in New Bedford, it was my good fortune to see a pretty near approach to freedom on the part of the coloured people.'

In the eighteenth and nineteenth centuries whaling and slavery co-existed as lucrative, exploitative, transoceanic industries; while whale-ships sought to disguise themselves as men o' war in order to forestall pirates (and sometimes harboured fugitive slaves themselves), slave ships seeking to evade Unionist blockades during the Civil War would masquerade as whale-ships. It was no coincidence that in 1850, as Melville began to write *Moby-Dick*, the issue of slavery was coming to a head. The stresses that would eventually sunder a nation also gave Melville's book its symbolic charge.

That year, a new Fugitive Slave Law gave owners extraordinary powers to pursue their 'property' over state limits. To America's great philosopher, Ralph Waldo Emerson, it was a 'filthy enactment'. Meanwhile, his Concord neighbour, Bronson Alcott – whose utopian, strictly vegan commune, Fruitlands, just outside Boston, was an early example of ethical living where the wearing of cotton was forbidden because it exploited slaves and where oil lamps were proscribed because they were the result of the death of whales – hid fugitives in a modern version of a Reformation priest-hole.

War between the states seemed imminent; and as the North and South argued over the right, or otherwise, to maintain their fellow man in chains, Melville turned the crisis into an elegant, cetological analogy.

> Some pretend to see a difference between the Greenland whale of the English and the right whale of the Americans. But they precisely agree in all their grand features; nor has there yet

> been presented a single determinate fact upon which to ground a radical distinction. It is by endless subdivisions based on the most inconclusive differences, that some departments of natural history become so repellingly intricate.

Elsewhere, Ishmael describes a whale of 'an Ethiopian hue', hunted until its heart burst; while the whiteness of Moby Dick itself seemed a reflection on America's preoccupation with colour.

Determined to protect his fellow fugitives from 'the bloodthirsty kidnapper', Frederick Douglass began an unprecedented campaign, the first black man in America publicly to oppose such injustice. Historians like to imagine that Douglass and Melville saw each other in New Bedford's narrow streets; in the same year that Melville sailed from the port, Douglass was 'discovered' lecturing on abolitionism in the Nantucket Athenæum. Four years later, the publication of his memoir, the *Narrative of Frederick Douglass*, attracted violent opposition. Some even questioned the author's authenticity, turning on his fierce beauty – not quite black, not quite white – calling Douglass a 'negro imposter' and 'only half a nigger' (to which he retorted, 'And so half-brother to yourselves'). In May 1850, Douglass's appearances in the New York Society Library – the same building in which Melville was even then researching his story of the White Whale – were disrupted by 'Captain' Isaiah Rynders and his Law and Order Party, a gang that attacked abolitionists, foreigners and blacks, encouraged by one newspaper which demanded its readers

STRIKE THE VILLAIN DEAD.

When Douglass strolled up Broadway with his two English friends, Julia and Elizabeth Griffiths, passers-by uttered exclamations 'as if startled by some terrible sight'. Worse still, when walking near the Battery, the trio were set upon by five or six men shouting foul language; Douglass was hit in the face, and the women struck on the head. It was a scene that had its counterpart in Melville's autobiographical *Redburn*, published the previous year, in which the young sailor sees his ship's black steward walking the Liverpool streets 'arm-in-arm with a good-looking English woman', and remarks: 'In New York, such a couple would have been mobbed in three minutes; and the steward would have been lucky to escape with whole limbs.'

Douglass reacted to these assaults in his essay, 'Colorphobia in New York!', and later became Abraham Lincoln's adviser on slavery during the bitter war that followed. Melville, whose father had been a friend of the Liverpool abolitionist William Roscoe, would invest *Moby-Dick* with the same blackness and whiteness, the same deceptively simple quandary. Strangely intertwined in history, slavery and whaling were both expressions of antebellum America; both doomed by their reliance on unsustainable resources, human and cetacean.

By the time Melville arrived, New Bedford was experiencing an unparalleled boom. In the 1840s, three hundred whale-ships – more than half of the American fleet – sailed from the port, often returning with two or three thousand barrels of oil and profits running into hundreds of thousands of dollars. Many New England boys, fired up by the heroism and glory it offered, volunteered for the chase. While their peers went to California in search of gold or the Dakota plains for buffalo, they found another wilderness: whaling was the Wild West of the sea.

Like a cowboy or a jockey, the experienced whaler was physically tailored for the job – or perhaps his job moulded him. 'He is a rather slender, middle-sized man, with a very sallow cheek, and hands tanned of a deep and enduring saffron color,' wrote Charles Nordhoff, who sailed from New Bedford soon after Melville, '. . . very round-shouldered, the effect possibly of much pulling at his oar.' A vagabond cast in 'this shabby part of a whaling voyage' – as Ishmael puts it – the well-travelled whaler bore

> a singular air of shabbiness . . . His shoes are rough and foxy, and the strings trail upon the ground, as he walks. His trowsers fail to connect, by several inches, showing a margin of coarse, grey woollen sock, intervening between their bottoms, and his shoes. A portion of his red flannel drawers is visible, above the waistband of his pantaloons; while a rusty black handkerchief at the throat, fastened by a large ring, made of the tooth of a sperm whale, and inlaid with mother-of-pearl, keeps together a shirt bosom . . . innocent of a single button.

The whaler was a kind of pirate-miner – an excavator of oceanic oil, stoking the furnace of the Industrial Revolution as much as any man digging coal out of the earth. Whale oil and whalebone were commodities for the Machine Age, and owners and captains adopted the same punitive practices employed in mills and factories, reducing pay and provisions to pursue a better profit.

When you're fast to a whale, running risk to your life
You're shingling his houses and dressing his wife.

It was to this often iniquitous trade that innocent young men found themselves signed up, almost unwittingly. Engaged in New York, they were shipped to New Bedford, the price of their passage deducted from the seventy-five dollars they had been promised. Sometimes the 'landsharks' got them drunk and virtually press-ganged their victims, who woke to find themselves aboard an outgoing ship, unable to get off.

At worst, whalers were treated like migrant workers, little better than bonded labourers. Nordhoff spent months 'in all the filth, moral and physical, of a whale-ship', and returned feeling that he had thrown away two years of his life: whaling, he declared, 'was an enormous, filthy humbug'. One young whaler came home after a five-year voyage to discover that while his friends had made their fortunes in the gold fields, he had earned just $400, half of which he owed in outfitter's bills.

To Ishmael, New Bedford was a 'queer place', a city that wore 'a garb of strangeness'. It certainly mystified Nordhoff when he first arrived. This whaling metropolis reached out from a corner of New England to light the world, yet it was remarkably still.

'One would never guess that he stood within the bounds of a city which ranks in commercial importance the seventh seaport in the Union, and whose ships float upon every ocean.' The reason for this complicit silence was the confinement of the port's commerce to a relatively small downtown area, as if it were keen to restrict its vulgar, even disreputable transactions to a whalish ghetto.

New Bedford is still a blue-collar place, a working port; perhaps that is why I like it so well: it reminds me of my home town. It still conducts its business from the same buildings used by the whaling trade; newspapers are published and radio stations broadcast in Portuguese; and in the north end of town, Antonio's restaurant sells salt cod and shrimp fritters to descendants of whalers and mill-workers on a Friday night. As the customers sit drinking at the bar, with an icy wind blowing down the street outside, it isn't hard to imagine a modern-day Ishmael walking in the door; or even his creator.

When he arrived in New Bedford on that bleak December day in 1840, Melville saw the city rising 'in terraces of streets, their ice-covered trees all glittering in the clear, cold air', endlessly unravelling in a panorama of activity. The port was alive to the business of the whale. Scores of ships lay in dock, preparing for long voyages, taking on supplies in holds which, as they emptied, would be filled by the fruit of their hunt. It was an efficient exchange: if 'greasy luck' was with the *Acushnet*, she would never need ballast. Flat-pack barrels – loose staves to be assembled by onboard coopers – provided further storage. Other ships were drying their sails like cormorants' out-held wings, their cargo unloaded by men back from tropical seas; they were easy to spot, for their sunburn glowed next to the pale faces of those who had wintered at home.

The wharfside was a centre of industry, like the whales never still by day or night, piled high with 'huge hills and mountains of casks on casks' while 'the world-wandering whale-ships lay silent and safely moored at last'. Here Ishmael listens to carpenters and coopers at work, 'blended noises of fires and forged to melt the pitch'. It is a Sisyphean sign, both quickening and deadening: 'that one most perilous and long voyage ended, only begins a second; and a second ended, only begins a third, and so on, for ever and for aye. Such is the endlessness, yea, the intolerableness of all earthly effort.' This was a task as dreary as a container ship sailing predetermined distances, bringing stuff in, taking stuff out, heavy with oil and whalebone and human effort.

Nordhoff too saw wharves laden with 'harpoons, lances, boatspades, and other implements for dealing death to leviathan'.

Beyond lay the inns and offices, chandlers and sail lofts, smithies and dining rooms, banks and brokers, all trading on the whale, directing every effort down to the river and the ocean beyond in an unremitting, profitable pursuit. Clapboarded and shingled, walled in wood like ships themselves, the five blocks tethered off Water Street – 'New Bedford's Wall Street' – were said to be the busiest in New England. This main thoroughfare, running uphill from the waterside, was devoted to the outfitters' shops and suppliers, while its side streets were home to boarding houses kept by whaling widows 'for numerous youthful aspirants to spouting honors'. For other honours, they could visit a waterborne brothel, anchored offshore.

Removed from this gritty business were the grand mansions of County Street, New Bedford's most prestigious address. These houses still occupy block after block in every permutation of architectural style, their details picked out in contrasting colours, each wildly different, yet each the product of factories that turned out decorative trim by the yard. Like the millionaires' 'summer cottages' in nearby Newport, Rhode Island, they vie with each other for extravagance. Most magnificent of all is the house built in 1834 for the whaling Quaker, William Rotch Junior, whose grandfather Joseph came from Nantucket to found New Bedford's industry.

Occupying a block of its own, this elaborate pile, with its verandahs and parterres, its reception rooms and bedrooms, seems incompatible with its owner's austere face, long silver hair and plain black coat. Nevertheless, William Rotch presided over the world's greatest whaling fleet from the glazed lantern that sits on the roof like a lighthouse, looking down on the waterfront and the source of his wealth. On a darkening winter's afternoon, I climbed to this eyrie through the attic-like servants' quarters, the sodium lights of the port already twinkling in the distance. 'Nowhere in all America will you find more patrician-like houses; parks and gardens more opulent, than in New Bedford,' Ishmael declares. 'Whence came they? how planted upon this once scraggy scoria of a country?' His answer lay with 'the iron emblematical harpoons round yonder lofty mansion . . . Yes; all these brave houses and flowery gardens came from the Atlantic, Pacific, and Indian oceans. One and all, they were harpooned and dragged hither from the bottom of the sea.' For every stoop and pillar on County Street, a whale died; each extravagance was bought at the cost of a cetacean. Oil for marble, baleen for wood, this was the rate of exchange from sea to shore.

And down at the quayside late at night, where the fishing fleet lies tethered to rusty piles, hulls bumping gently and engines purring, I wonder how it must have been for these young men to ship out from this port, to leave these homely waters for uncertain seas. A sense of utter abandonment to fate, disconnecting from America, seeking escape wandering the oceans, orphans in search of a new home among a family of men, yet enslaved to the movements of the whale, man and animal forever linked.

The next morning, as I leave, snow starts to fall, turning the mural over the highway into an impressionist canvas, flecked with white. As the traffic picks up speed, I look over my shoulder. The painted whales are fading from view, losing their shapes. A hundred yards more and they are gone, vanishing with the city into the flurrying swirl, to be replaced by the concrete clamour of the road ahead.

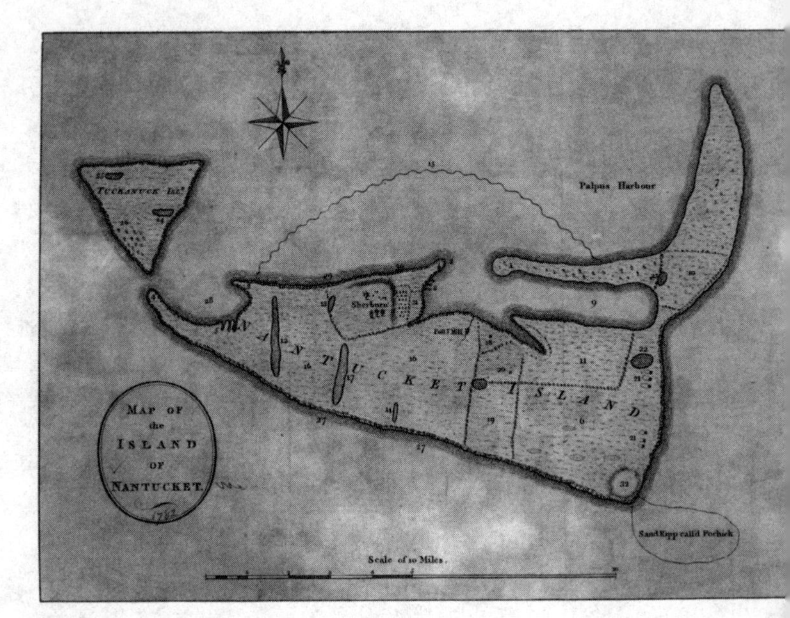

MAP OF
the
ISLAND
OF
NANTUCKET.
Tuckanuck Isl^d.
Palpus Harbour
Sherborn
NANTUCKET ISLAND
Sand Kepp call'd Pochick
Scale of 10 Miles.

V

Far Away Land

> Nantucket! Take out your map and look at it. See what a real corner of the world it occupies; how it stands there, away off shore, more lonely than the Eddystone lighthouse . . . a mere hillock, and elbow of land; all beach, without a background . . . What wonder, then, that these Nantucketers, born on a beach, should take to the sea for a livelihood!
>
> Nantucket, *Moby-Dick*

Off-season Hyannis is deserted, closed for the winter. This morning's storm has cancelled ferry sailings; the evening's schedule may be called off as well, the seas too high for a safe crossing. It seems that, like Ishmael, I will be frustrated in my attempt to make Nantucket tonight. It is the coldest weather of the year, and the wind is picking up. In the ferry office, the woman delivers the expected news. But what about the plane? she says. There's fifteen minutes to the last flight.

On the darkened runway, the light aircraft rumbles along until its wings seem to stretch and straighten. Soon the sodium flares of the town fall away, to be replaced by the silver-black waves far below. I'm sitting in the co-pilot's seat; the young pilot wears a baseball cap, and the cockpit smells of his sandwiches. The dual controls tick and turn in my lap. Through the windscreen I see a shape on the horizon, bracketed by flashing lighthouses. A clutter of stars bursts around Orion. Twenty minutes later we are falling through the clouds, twin beams meeting in the mist, guiding us down. With a bump the tyres bite tarmac, and as we few passengers step out onto the airstrip, Flint the boxer dog scents home.

When Ishmael and Queequeg arrive in Nantucket, by schooner from New Bedford, they put up at another inn while they search for a suitable whale-ship. As they do so, Ishmael takes the opportunity to delineate the island in great detail, from its remarkable history down to its clam chowder – even though his creator had never actually been there. Such was Nantucket's fame: it already lived in the American imagination, a name that summed up the pioneering, heroic spirit of the new republic. Early cartographers even saw the shape of a whale in its harbour, as if its myth were incarnate in the island's very geography. But like its neighbour, Cape Cod, Nantucket was both part of America, and set apart from it at the same time.

The word is Native American, *Nattick*, meaning far away land; and from far away, its wharves once stank so much that visitors could smell the island before they saw it. Now they bob with expensive boats gleaming with brass and veneer. The town's Main Street is unevenly paved with hefty stone setts, undulating as if to shrug itself of unwanted visitors. Smart shops and old-fashioned drugstores with high counters serving sodas and sandwiches give way to sandy lanes lined with clapboard homes. Many have door-knockers and weathervanes in the shape of whales, 'but they are so elevated, and besides that are to all intents and purposes so labelled with "*Hands off!*" as Ishmael complains, 'you cannot examine them closely enough to decide upon their merit.' Nearby is the Athenæum where, in 1841, Frederick Douglass spoke to a mixed-race audience at the island's first anti-slavery convention; a second meeting the following year ended in a riot. It would be hard to imagine such insurrection here nowadays.

The higher up the hill you go, the more the houses increase in size. Unlike New Bedford's showy homes, however, they announce their wealth quite quietly. Three identical buildings, built in the 1830s by Joseph Starbuck for his three sons, were the first brick houses on the island; they speak of a fantastical New England. Even a century ago Mary Heaton Vorse saw Nantucket as 'some beautiful old woman sitting dreaming in a garden . . . proud of her faded and excellent beauty'; its summer visitors already outnumbered year-rounders, and 'no immigrants swarmed through the wide houses of the old whaling captains, as in New Bedford'.

Nowadays, an island which furnished the world with the names of Macy, Folger and Starbuck rejects commerce. There are no supermarkets selling cheap postcards, no homeboys' stores with piles of jeans. It all adds up to a faintly unreal perfection. The cold light turns each streetscape into an exquisite composition of towers and trees laid bare by an acid-blue sky. Colours shade into each other; flat grey shingle and dusty green lichen; roots disrupt brick pavements with slow-motion earthquakes.

These lanes also lead back to one place. New Bedford's mansions were dragged up from the ocean; these houses were landed at the harbour in barrels; came totalled in copperplate figures in bound books; were marked in ivory teeth over years spent on the other side of the world. They may look innocent, but they too were built by heathens and monsters.

In Nantucket's refurbished whaling museum, the bones of a sperm whale look across to a wall arrayed with harpoons and lances like medieval hardware in the Tower of London. Upstairs, the galleries are filled with the more delicate by-products of this bloody business. Standing on plate-glass shelves are fine examples of scrimshaw, a craft which in itself was an expression of an industry of excess.

On long voyages, the large crews required to hunt whales were idle for much of the time. To occupy hands that might be otherwise engaged, they were given whale teeth on which to record images of their fancy or everyday life. Soaked in brine to preserve their suppleness and polished with sharkskin, the teeth – which could be up to ten inches long – were etched with needles or knives, creating patterns to be inked with soot from the ship's try-pots. Some were little more than graffiti; others were traced with illustrations snipped from Victorian periodicals, or imaginary classical scenes. Often, they portrayed the ships themselves.

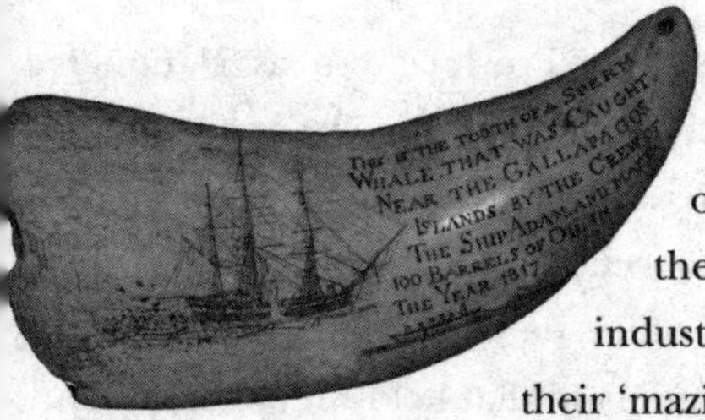

Decorated with bosomy women or fey-looking youths or feats of whaling endeavour, these were folk artefacts of an industrial age. Ishmael compared their 'maziness of design . . . full of barbaric spirit and suggestiveness', to engravings by 'that fine old savage, Albert Durer'. Tactile lumps of creamy-smooth ivory once gripped in a seaman's fist, they are imbued with a sensual, primitive significance akin to tattooing, 'or *pricking*, as it is called in a man-of-war'. As their designs resembled tattoos on a sailor's biceps, so tattooing instruments, themselves distinctly tribal, were made with whale-ivory handles, while other sailors assembled 'little boxes of dentistical-looking implements', custom-made for scrimshanding. They were direct records of the whalers' experiences and desires, journals for illiterate men. Some were decorated with pornographic cartoons, or were carved into phalluses.

The most artful pieces mark the pomp of whaling; the peak years of scrimshaw were those of the great voyages to the South Seas in the 1830s and 1840s, when whale bone was also turned into delicate 'flights'– trellis-like structures for winding yarn – or carved into pastry-cutters to be sold in fancy-goods stores or given to loved ones. But as history moved on, these macabre objects languished in attics, unloved, unvalued; only in the late twentieth century were they seen anew, and one man in particular was responsible for their revival: John Fitzgerald Kennedy.

The Kennedys are synonymous with the Cape and its islands, an American aristocracy convened around the family compound at Hyannis. Even before he became thirty-fifth President of the United States, John F. Kennedy had moved to declare the Outer

Cape's beaches from Eastham to Provincetown as National Seashore, sacrosanct from urbanization. And it was as an extension of his love of maritime New England that Kennedy began to collect scrimshaw. Soon his collection stood at thirty-four whale teeth, favourite examples of which he kept on his Oval Office desk, to be turned in the same hand that held the world in its balance.

In 1963 the First Lady ordered a special Christmas gift for her husband – a whale tooth engraved with the presidential seal. He would never receive it. Shortly before he died, the President threw a private dinner for Greta Garbo at the White House, when he gave the actress a piece of scrimshaw. 'I might believe it a dream,' Garbo wrote to Mrs Kennedy afterwards, 'if I did not have in my possession the President's "tooth" before me.' Two weeks later, on the night before his funeral, his widow placed her husband's Christmas present in his coffin. It was a potent act: the king of Camelot interred with the talisman of a heroic age; a relic invested with the power of its original owner. It was a ritual as charged as Ishmael's claim that the British monarch was anointed with whale oil –

> Think of that, ye loyal Britons! we whalemen supply your kings and queens with coronation stuff!

– while the President's amulet was ready for the moment when this Arthur was needed anew; as if he might yet scan the Atlantic horizon with his pale blue eyes, waiting for the whales to reappear.

It was on Nantucket that modern whaling began; on its narrow shoulders lies the glory. In 1659 nine new citizens acquired the rights to the island, Quakers such as Thomas Macy, Tristram Coffin and Christopher Hussey who had suffered

Puritan persecution in New England. For an island that 'seemed to have been inhabited merely to prove what mankind can do', whaling came as a kind of destiny, as told in Obed Macy's *History of Nantucket*, and quoted by *Moby-Dick*'s sub-sub-librarian:

> In the year 1690 some persons were on a high hill observing the whales spouting and sporting with each other, when one observed; there – pointing to the sea – is a green pasture where our children's grand-children will go for bread.

For centuries Nattick Indians had foraged for whales in these plentiful waters. The new Nantucketers learned from their techniques. At first, land masts with crude ladders were used to spot right whales on their migration north. Harpooned and towed back to the beach, their two-foot-thick blubber yielded more oil than any other whale, and their baleen was taller and finer – the same 'limber black bone' from which Captain Peleg's wigwam is constructed on the deck of the *Pequod.*

Then, in 1712, a new prey was discovered. According to legend, Christopher Hussey was out hunting when his sloop was blown beyond the normal limits of the Nantucket fisheries. There, in deep waters, he encountered the sperm whale, hitherto considered 'fabulous or utterly unknown', says Ishmael. Now it would usurp the right whale 'upon the throne of the seas'. In assuming that crown, the sperm whale became a more fitting quarry for the lordly islanders, 'this horrible and indecent Right whaling, I say, compared to a spirited hunt for the gentlemanly Cachalot'. Whaling for such a noble animal was like riding with foxhounds compared to the lowly bear-baiting of right whales. Soon it was a crucial part of the island's economy, still more so as the right whales became scarcer. By 1730 there were twenty-five vessels in the island's fleet. By the end of the century, it would lead the world in whaling.

'And pray, sir, what in the world is equal to it?' quotes our sub-sub-librarian, from *Edmund Burke's reference in Parliament to the Nantucket Whale Fishery*. Burke went on to inform Britain of 'the progress of their victorious industry': 'No sea but what is vexed by their fisheries. No climate that is not witness to their toils.' Old Europe could not rival 'this recent people, a people who are still, as it were but in the gristle and not yet hardened into the bone of manhood'. The new nation seemed to prove itself by the whale. For Owen Chase, first mate of the *Essex* and scion of an old Nantucket family, he and his fellow whalers were crusaders 'carrying on an exterminating warfare against those leviathans of the deep'. They were knights and squires bound up in a new chivalric order; outriders of an empire, even as the whales were driven 'like the beasts of the forest, before the march of civilisation into remote and more unfrequented seas'.

It was a pattern of plunder of the New World's resources. As their land-borne counterparts drove buffalo from sixty million to extinction, so these oceanic cowboys pursued whales to the brink. It was as if the antediluvian beasts had to die in order to assert the modern world. For America, the common enemy was the wilderness; and just as that wilderness was in fact full of animals – and native peoples – so the American seas were full of whales, ready for the slaughter. Hostilities were declared in 1712; it would be a war of attrition ever after.

At first Nantucket whaling was a family business, a trade passed down from hand to hand. Any young man of promise could expect, after two whaling trips, to captain his own ship. Crews were 'composed of the sons and connections of the most respectable families on the island', Owen Chase wrote; 'they labor not only for their temporary subsistence, but they have an ambition and pride among them which seeks after distinguishment and promotion.'

Initially, whales were brought back to port to be rendered, but by 1750 shipboard try-works – a Basque invention of brick ovens with giant cauldrons in which to boil down the blubber – were being used. In a neat flip of cause and effect, these contraptions permitted the rendition of whales on the ever longer voyages required to find them. At the same time, whaling became part of a greater, political game. The Wars of Independence stalled Nantucket's growth – its fleet declined from one hundred and fifty to thirty-five ships – while the islanders attempted to remain loyal to Britain, their greatest customer. But with the new republic, the ships returned in greater numbers than ever.

> And thus have these naked Nantucketers, these sea hermits . . . overrun and conquered the watery world . . . let the English overswarm all India, and hang out their blazing banner from the sun; two thirds of this terraqueous globe are the Nantucketer's. For the sea is his; he owns it, as Emperors own empires; other seamen having but a right of way through it . . . There is his home; there lies his business . . .

In 1944 Ishmael's hymn to Nantucket was broadcast to American troops overseas as a means of raising morale, reminding them of an heroic age. 'Indeed a Nantucket man is on all occasions fully sensible of the honour and merit of his profession,' Owen Chase had written a century earlier, 'no doubt because he knows that his laurels, like the soldier's, are plucked from the brink of danger.' Here was honour untainted by 'the luxuries of a foreign trade'. Its reward was God's bounty for His own country.

Nantucket was the purest expression of this holy quest. Its houses seem to say it, plain and angled and sharpened against the light, as much ships as they were homes, their shuttered windows and narrow doors facing all fortune and affliction. New England ports sent out more ships a week than old England did in a year, and 'our sails now almost whiten the distant confines of the Pacific', boasted Chase. Through whaling, America reached across the world for the first time; whaling exported its culture and ideas. And Nantucket was at its heart. By 1833, seventy thousand souls and seventy million dollars were tied up in whaling and its associated crafts; ten years later that figure had nearly doubled. The United States exported a million gallons of oil to Europe each year. At its peak, no fewer than thirty-eight American ports would pit themselves against the whale, from Wiscasset in Maine to Wilmington in Delaware, although many failed in the attempt.

The appeal of this filthy business was money: vast sums of it, for some. An owner could expect a threefold return on his investment. The first industrial fortunes in America were built on the whale fishery. In New England it remained an industry controlled by the Quakers, who saw no contradiction between their pacifist beliefs and their daily business. It certainly did not concern Captain Bildad, part owner of the *Pequod*, who 'though a sworn foe to human bloodshed, yet had he in his straight-bodied coat, spilled tuns and tuns of leviathan gore . . . very probably he had long since come to the sage and sensible conclusion that a man's religion is one thing, and this practical world quite another.'

Of all the products that were made from the whale, the pure-burning candles produced on Nantucket were the finest, as if the Quakers' Inner Light shone from the whale itself. The technique of turning whales into candles was introduced to New England in 1748, by a Sephardic Portuguese Jew, Jacob Rodriques Rivera.

It was a complex process. Headmatter from sperm whales was brought directly from the ships to large wooden manufactories, where it was heated in great kettles to remove water and impurities. It was stored in casks and, over winter, cooled to a coagulate mass. This was placed in woollen bags, which were then compacted in a wooden press from which spermaceti trickled, like juice from an apple press, or oil from olives. This first pressing, the purest, was known as 'winter-strained' sperm.

The remaining matter was made into 'black cakes', and stored until the spring, when in warmer temperatures it began to ooze. Re-pressed, this was 'spring-strained' oil. After a third and final pressing, a brownish mess remained; heated with wood shavings and potash, it clarified like butter, and the result made pure white wax. It also made fortunes.

Kezia Coffin was scion of one of Nantucket's first families, a 'she-merchant' famous for her fine clothes, the forbidden spinet she played, and the opium she was reputed to use. She began selling pins, but her merchandising business expanded into whale products. Loyalist Nantucket continued to trade with Britain, and during the Revolution Kezia made a private deal with a British admiral to ship oil and candles to London, along with smuggled goods sold at inflated prices. Kezia was a paradigm of feminine fortitude and enterprise on an island of women used to the absence of men. 'Aye and yes, Starbuck,' as Ahab confesses to his first mate, 'out of those forty years I have not spent three ashore . . . leaving but one dent in my marriage pillow.' Whaling separated sexes; and in this isolated place, as isolated as any ship, and yet bleaker in midwinter, whaling 'widows' had recourse to opium to cope with the loneliness. Others used plaster dildos known as 'he's-at-homes'.

The American war with Britain complicated matters for Nantucket whalers. The island was officially neutral – not least because of the pacifism of its inhabitants. They were only allowed to sail from New England if they proclaimed themselves on the side of the rebels; but if they did, the British would claim their whale-ships. Some moved to Newfoundland or Canada to pursue their trade. Others sailed to the Falkland Islands to exploit its newly discovered whaling grounds on behalf of the British.

In the aftermath of revolution, Nantucket grew richer than ever on the wealth of whales. It also exported its trade and expertise. Nantucket Quakers had founded a whaling port at Hudson, New York, where, despite being a hundred and twenty miles from the sea, a thirty-five-strong fleet prospered. Other colonies were founded in Dartmouth, Nova Scotia, by Timothy Folger and Samuel Starbuck, and in 1785 Starbuck, Folger and William Rotch Senior made approaches to Britain about setting up a whaling port there. Rotch and his son Benjamin travelled to London for talks with the Prime Minister, William Pitt. After lengthy negotiations – Rotch wanted £20,000 removal costs and naturalization for thirty ships and five hundred of his countrymen, but during the talks Rotch set up at Dunkirk, having been offered better conditions by the French – the British finally invited the Nantucketers to create a new station at Milford Haven in 1792, granting them 'the rights and privileges of natural-born subjects'. Here, in a pre-echo of the Welsh who would settle in Patagonia, an enclave of Nantucketers was founded, complete with New England architecture, a Quaker meeting house, and a Pembrokeshire cemetery populated with Starbucks and Folgers.

Like other religions, Quakerism owed its power to its restrictions. Forbidden by their beliefs from swearing oaths of office, Quakers were debarred from professions such as the law and medicine. This had the effect of directing their talents into business, at which they succeeded pre-eminently. And while Quaker ethics also precluded the flaunting of wealth, they did allow fine materials to be used in simple designs; hence the unadorned architecture of Nantucket's 'Golden Age', an æsthetic that still shapes the island today.

Such wealth stood in sharp contrast to the growing black population that serviced it – initially slaves, then, with the Quakers' early abolition of slavery in 1773, free men and women. Some prospered in their own right: in 1822 Absalom F. Boston sailed on the *Industry* with an all-black crew, returning as the island's richest African-American, his success explicit in the thick gold earrings he wore in each ear. None the less, the island's ruling class remained resolutely white, reiterated in a roll-call of industrious names: Coffin, Chase, Folger, Gardner, Macy, Starbuck, Hussey. The street maps show house after house of them, a freemasonry of spermaceti; a territory divided between families and manufactories on an island-whale made out of whales, telled and ledgered and decanted from barrels and beaten into silver, the only precious metal acceptable to a Quaker.

Nantucket's skyline announced its own fortune. It was spiky with ship's masts, studded with lantern towers topped with whale-shaped weathervanes, and animated by windmills with cart-wheel props which gave the appearance of 'huge wounded birds, trailing a wing or a leg'. This little island was one big machine: processing whales and wind to create candles and flour. Stern, stalwart and blessed, Nantucket was a nation of its own, existing in the hearts of its men at sea and in the work of its women at home.

> For years he knows not the land; so that when he comes to it at last, it smells like another world, more strangely than the moon would to an Earthsman. With the landless gull, that at sunset folds her wings and is rocked to sleep between billow; so at

> nightfall, the Nantucketer, out of sight of land, furls his sails, and lays him to his rest, while under his very pillow rush herds of walruses and whales.
>
> Nantucket, *Moby-Dick*

But in the 1840s a succession of adversities began to turn Nantucket's fortune into failure. The new, larger whale-ships required to sail further for sperm oil could not negotiate the treacherous sandbar across the island's harbour, which had begun to silt up. Business began to favour the easier access of New Bedford, as did many islanders, who emigrated there. The new port was the brash newcomer; and while Nantucket's haughty sailors stubbornly pursued now depleted old hunting grounds, New Bedford's whippy young whalers profitably exploited the Pacific seas.

In 1846 a great fire destroyed a third of the town's businesses – burning all the brighter for warehouses filled with barrels of whale oil. Two years later, the Gold Rush tempted young Nantucketers in search of quicker fortunes. In 1849 the fittingly named *Aurora* was the first Nantucket ship to sail for San Francisco, where whale-ships lay abandoned as their crews deserted for the gold fields, joining the crowds flocking to the west; many left home with little or nothing, not even underwear, reasoning that they had gone to wash gold, not their own dirty linen.

The final knell for Nantucket came with other discoveries from the earth. From the 1840s, kerosene and coal gas were already lighting city streets and houses, although initially the use of domestic gas only encouraged the demand for whale oil as the passion for bright light spread. Then, in 1859, Edwin L. Drake drilled for oil on a farm in Titusville, Pennsylvania; the black-gold spurt that gushed from the ground like a whale's spout signalled the end of the sperm whale fishery – and the beginning of another elemental plunder.

After fire and oil came war. Four hundred Nantucket men and boys left to fight the cause of the Union, as Confederate ships wreaked havoc on the Yankee whaling fleet. Many ships were captured or burned, causing other owners to keep theirs at home. Some were sacrificed by the Union itself: forty whale-ships – known as the Stone Fleet – were filled with rubble and scuttled to block southern harbours. The industry limped on for a few more years, but in 1869 the last whaling ship left Nantucket.

Slowly, surely, the island was cut off from time. Sealed from the modern world like land requisitioned by the military, its blasted heaths remained pristine, its cottages hidden in hollows away from fierce Atlantic winds. Cobbled streets fell silent, unclattered by carts carrying barrels of oil. Blank windows of brick mansions built by Quaker captains looked down to an empty quayside, while their owners lay in barren graves.

"THERE SHE BLOWS!"

VI

Sealed Orders

> *Wm. Bartley.* How came you to think of running away? Why sir, to tell you the truth I am afraid of a whale . . .
>
> Examination of deserters from the whaling ship, *Houqua*, 1835

Down the coast in Connecticut, white clapboard houses rise out of the hoary grass like Christmas cakes. At dawn, every puddle has turned to ice; even the moss cracks beneath my feet. According to my hosts, this road is one of the oldest in New England, an Indian trail turned into a colonial way. Last night, as I walked by moonlight along the deserted lane, I imagined shapes at the dark edges where house lights yielded to the woods and civilization abruptly fell away.

This morning, the sun climbs over granite rocks, and the highway that crosses the lane is already roaring with trucks. On the other side is the river, widening towards the sea and the site of another whaling port: Mystic. This, too, is a place of memory. Here, in 1637, the Puritans waged war on the Pequots, killing four hundred men, women and children. Perhaps it was no coincidence that Ahab's ship appropriated the name of this slaughtered tribe. Or that, looming through the leafless trees ahead of me are the masts of the *Charles W. Morgan*, America's last remaining whale-ship, built and launched on the Acushnet in the same year that Melville sailed on the voyage that would inspire the story of the *Pequod*.

But the *Morgan* is no fantastical vessel with a whale jaw for a tiller or whale teeth for pins. This is a real ship with all its constrictions and discomforts; an instrument stripped to its essential parts. Everything here was designed for the collection, production and storage of the whale, rather than the comfort of those expected to process it. This was a mobile factory, a nineteenth-century oil tanker; but it is also surprisingly sleek, like the clippers that ferried tea to England from Ceylon, and on one of which my own ancestor was a captain until he was lost at sea.

The *Morgan* is laden with equipment, almost top-heavy with it. As I duck through shrouds and step over holds, I am aware of how much danger such a ship represented for the unwitting, even before it set sail. Swaying ropes and blocks meant every movement

had to be made with care. Here life was lived in public; even the captain shared his stateroom, a semblance of a landlubber's salon, dining room and study all shrunk into one. The space seems homely, with a faded red sofa built into the side like a bunk in a caravan. In the captain's cabin itself, an ornate bed is gimballed so as to swing in high seas and rock its occupant to sleep; and in the corner, a cupboard conceals the only private 'head' onboard.

In this miniature world – so small compared to the ocean all around – every inch is used efficiently. Shelves fit into corners, drawers are set above the sofa, chests stowed under bunks. Lamps hang from hooks, pots and pans stand in compartments to stop them rolling around the galley – itself little more than a larder. There is a neatness worthy of a Shaker interior; a cosy arrangement, like a grown-up Wendy house. Sometimes an entire family travelled in these quarters. Through their eyes I see life lived on board, children at their schoolwork on the table built around the mast, their mother sewing as the ship lurched to and fro. One four-year-old, Eugene, playing in a whaleboat, nearly fell overboard, screaming for his Pa as he clung to the side. At bedtime their father told them stories about what the whale said and did.

The reality of ship life was less comforting. There are cupboard-like cabins for officers and mates, the accommodation growing ever smaller as rank reduces until, beyond the blubber room, double tiers of bunks are built into the narrow forecastle, shelves for human stowage. Here the lowly slept, clustered like cockroaches at the prow, subject to class distinction even in the light they were allowed. Set flush into the deck are solid glass prisms, shaped like upturned hexagonal pyramids – so-called deadlights that could concentrate the sun's rays, producing a luminescence equivalent to seventy watts. But theirs was an undemocratic illumination: while the staterooms boasted a cluster of these nineteenth-century bulbs, the forecastle had just two, shedding a watery light barely enough for a sailor to read in his bunk; and that was frustrated when obscured by a stray rope on the deck above.

The forecastle was seldom a good place to be. One sailor claimed to have seen 'Kentucky pig-sties not half so filthy, and in every respect preferable to this miserable hole'. Not only was it dark and odorous, but damp, too; in bad weather, crewmen might spend days on end wearing wet clothes. 'Those who had been to sea before found this nothing new,' wrote Nelson Cole Haley, aged just twelve years old when he ran away from his home in Maine. Now sixteen, Haley signed on for the *Morgan*'s voyage of 1849–53 as a boat-steerer.

> Still, it was hard, even for them. After standing their watch, often wet through as soon as they came out of the forecastle, they had no chance to change clothing, if they had dry to put on, until they were relieved and went below. There twenty-five men lived in quarters so small that it was impossible for all of them to find standing room at one time . . . And this was not for one day or month, but was their only home for four years . . .

In the tropics, the unrelenting sun which sparked deadlights into life made duties even more difficult to bear. As the ship languished in dead calm, with no whales seen for weeks, lassitude overtook the company. The top deck was kept cool by watering, and its walking larder of pigs squealed with delight when buckets of sea water were thrown over them, too. Some men could shelter in the shade of the sails, but those aloft 'had to take it straight up and down', their eyes dazed by the cataractic sun reflecting off the sea. In the forecastle, the heat was worse. 'The watch below lay in their berths trying to sleep, the perspiration streaming from their bodies, with nothing but the curtains drawn in front of their bunks for covering.'

Yet even on such a ship it was possible to keep secrets. One welcome relief for sailors was a gam, a meeting with another vessel when letters and news were exchanged and men could socialize. On one such gam with the *Christopher Mitchell* of Nantucket – the same ship whose previous captain, William Swain, had been lost to a whale, as his memorial in the Seamen's Bethel testified – young Haley heard about one of her crew who, despite 'showing no more fear of a whale than the bravest of green hands', had faced jibes about his appearance. On falling ill in his bunk, was seen naked – and found to be a woman.

This anonymous Orlando told an extraordinary story. Her lover had promised marriage, only to run away to sea. Through the services of a New York detective, she discovered he had signed up to a whale-ship. Not knowing which one, she set off for New Bedford, where she bound her breasts with calico and, being tall and slim, 'passed herself off as a green boy who wanted to go a-whaling'. After her confession, she broke down in tears, but was comforted by the captain, who found her rather attractive, once she had sewn herself a loose dress, and as sickness and shade returned to a more lady-like pallor. When the ship called at Lima, the woman was placed in the hands of the American consul; only when the *Christopher Mitchell* returned home did her story become public.

Such was the no-man's-land of the whale-ship, where boys would be boys and girls would be boys, too. Life onboard was peculiar to itself and of itself: enclosed yet open, confined yet free, disciplined yet liberated. For months on end a ship's crew knew only this world. Time was measured in the watches of the day and by the shadows of the masts; on the featureless ocean they might be anywhere on earth, living within wooden walls, a colony of men ruled over by erratic officers and determined by the wilful meanderings of whales. Yet for all the depredations, the romance remained. Why else would men volunteer for this life, if not for its sense of adventure? Hardly for the pay, or the conditions.

It was this containedness about which Melville wrote so well in his novels of the sea, especially in the two works that preceded *Moby-Dick*: *Redburn*, a fictionalized account of his first sea voyage to Liverpool; and *White-Jacket*, another slice of his life story whose subtitle proclaims 'The World in a Man-of-War'. It is set on board a naval ship, a 'bit of terra firma cut off from the main; it is a state in itself; and the captain is its king . . . Only the moon and stars are beyond his jurisdiction.' Here men lived 'in a space so contracted that they can hardly so much as move but they touch ... the inmates of a frigate are thrown upon themselves and each other, and all their ponderings are introspective.'

Such intimacy permitted desires forbidden by the civilized world. Redburn extols the beauty of his English shipmate Harry, with dark curling hair and 'silken muscles', and a complexion as 'feminine as a girl's'; an equally handsome Italian boy plays his concertina with a suggestive enthusiasm almost embarrassing to read. The narrator of *White-Jacket* is more circumspect, although he notes that one midshipman is 'apt to indulge at times in undignified familiarities with some of the men'. When they resist, he has them flogged – a scenario that would inspire Melville's last work, *Billy Budd*, in which the villainous first mate, Claggart, becomes obsessed with the Handsome Sailor, Billy or Baby Budd, with fatal consequences for them both. In real life, other seamen found different outlets: Philip C. Van Buskirk, a contemporary of Melville's, left a startlingly frank journal of his onboard addiction to self-abuse.

Ishmael himself is never more than ambiguous on such matters; but since nothing in his creator's work is accidental (about the only thing his critics agree on), it is impossible not to see a pattern in Melville's emblematic titles –

RED BURN
WHITE JACKET
MOBY DICK
BILLY BUDD

– books set in a world without women and in an age that had no name for love between men (although his peer, Walt Whitman, devised the term 'adhesiveness' for what he felt for his fellow man). From the fiery youth of *Redburn* to the masculine discipline of *White-Jacket*, from the phallic pallor of *Moby-Dick* to the virginal *Billy Budd*, Melville fictionalized his past and obscured his emotions in a matrix of literary intent.

The sea was the perfect arena for such arch invention. A fatherless middle-class boy had deliberately placed himself as far from land – and female influence – as it was possible to be, creating a new family, and a new identity for himself. Instead of his mother and his sisters, he answered to a captain and lived among men. Removed from the security of home, and freed from its confines, Melville was launched into the brutal reality of living with men united only in the common pursuit of a bloody business. He and his fellow sailors had cut all ties with civilization, sailing to islands where murderous natives with filed teeth threatened to eat their shipmates. They were boys in a boy's own story, although they travelled on a vessel whose very ceilings kept them down, as if perpetually tugging their caps.

Descending below the waterline to the *Morgan*'s hold, I feel as though I were within the whale, contained within wooden ribs. In the dampness, I sense the pressure of water from without, even knowing this massive chamber is braced by sturdy knees cut from live oak, like flying buttresses on a great cathedral. The church-like air is illusory, for this maritime crypt was filled with barrels of oil, a visible measure of success, an ascending scale marked up the hull as a prisoner ticks off the days on his cell wall. It was in everyone's interests – from the captain with his magnified lay, to the seaman's humble fraction – to see this space diminish. Each barrel represented incremental profit; its absence, potential loss.

The *Morgan*'s timbers are still stained with decades of oil. Like the candle works of Nantucket, whose infused floorboards oozed when they were removed, the years have left this vessel saturated with the products of the animals she had processed. As a whale's skeleton retains its sap, so these soaked knees and ribs became the bones of her prey, transforming this death ship – this whale widow-maker — into a simulacrum of the creatures she pursued. In 1941, when she was brought to Mystic for restoration, objects were found between the *Morgan*'s bilges: bits of clay pipe, coins, whale's teeth, and strange shell-like bones – the inner ears of a whale – archæological relics that had rattled around for decades in the belly of the vessel. It was as if the ship had become a repository of herself.

Back in the staterooms, sitting at the captain's table as the wind sways the ship to and fro on her moorings, breaking the ice around the bows which promptly refreezes into abstract shards, I try to imagine life lived in this wooden box filled with more than forty men and boys and the rendered fat of tens of whales. Perhaps such conditions merely merged men into the visceral business in which they were engaged; perhaps they gave up their humanity for the duration, to wallow in whale oil for its own sake; to live and die for the whale.

Melville sailed on the *Acushnet* from New Bedford on Sunday, 3 January 1841. He may have been no greenhand – despite what it said on his shipping papers – but his earlier passage to Liverpool, carrying cotton rather than oil, bore little resemblance to the adventure that lay ahead of him.

Once at sea, the mates made their selection for the whaleboat crews. Mustered aft, men were interrogated about their experience as the mates checked their hands and feet and felt their muscles in an inspection that resembled a slave auction. Ships had three or four such crews, comprising the captain, or a mate, four foremast hands (as Melville was), and a harpooneer; fewer than five men might be left behind to run the vessel when the boats were lowered from the divots on which they hung on the ship's side, ready for action. As with everything in whaling, periods of frenetic energy alternated with soporific inaction or numbing drudgery. Time itself was different at sea. Far from land, the levelling ocean flattened out the days to be recreated in nautical dispensations, reordered from noon to noon.

First part, noon to 8 pm
Middle part, 8 pm to 4 am
Latter part, 4 am to noon

Four hours on, four hours off, *watch and watch* regulated the crew's life. When no whales were seen, the ship would sail in and out of as yet undetermined time zones. When the chase was on, time would accelerate, or even disappear. And all this – all these men, all their efforts, all their aspirations – existed for those few minutes when a whale might be won. All this human striving – from recruitment and requisition to searching and finding a distant disruption, followed by the frenetic hunt – all in order to fill wooden barrels that would ensure only a brief stay on land till the call to sea came again. As Ishmael says, the whole process was a remorseless cycle; a man might not be free of it till nature or the whale's caprice released him. As surely as those shrouds held the mast to his ship, as surely as the line held the harpoon to the whale, so was the sailor tethered to his prey in an unerring deposition of faith.

'Ah the world! Oh the world!'

Decks were scrubbed, men sent aloft for two-hour watches to look for whales. Until then, the ship and her crew lay in a kind of limbo. New recruits would practise in the boats, hardening muscles and honing co-ordination in a gymnasium of the sea. They rehearsed their techniques on passing porpoises or pilot whales, whose oil would occasionally be mixed with spermaceti to swell the profits of less honest whalers. For sixty-nine days the *Acushnet* sailed on a course now unknown to us, although it is probable that, like most New England whalers, she called at the Azores, seeking fresh provisions and new hands. Only then, as they sailed over the five-mile-deep chasms of the mid-Atlantic, would the hunt begin in earnest.

Sperm whales are not bound to seasonal migrations like the humpbacks; even so, they still roam tens of thousands of miles each year, often congregating in certain areas known as 'grounds' to the whalers. These were plotted on mariners' charts, marked with whalish symbols like maps in a military campaign. A favourite ground was the equatorial region, the Line. Here, at the earth's midriff, the whales seemed to gather as if in a preordained meeting with their fate.

The men had watched for weeks from the topgallant crosstrees, tiny figures swaying ninety feet in the air, everyone waiting for the magic words –

There she blows! T-h-e-r-e s-h-e b-l-o-w-s!

– at which the animals would appear, as if mystically summoned from the deep.

> And lo! close under our lee, not forty fathoms off, a gigantic Sperm Whale lay rolling in the water like the capsized hull of a frigate, his broad, glossy back, of an Ethiopian hue, glistening in the sun's rays like a mirror.

Sometimes they saw twenty or thirty whales riding the waves like surfers, 'tumbling about when the big seas would catch them and almost turn them over', as the teenaged Haley recorded, with not a little admiration. 'Sometimes one could be seen on the crest of a wave. As it broke he would shoot down its side with such a speed a streak of white could be seen in the wake he made through the water. When reaching the hollow between two seas he would lazily shove his spout holes above the water and blow out his spout, as much as to say, "See how that is done."' But even as the young whales were at their sport, the order was given to lower the boats.

The Yankee whaleboat was 'the most perfect water craft that has ever floated': a sleek, sharp-pointed vessel thirty feet long, yet 'so slight', as Melville wrote, 'that three men might walk off with it'. Double-ended for maximum manœuvrability, enabling it to be rowed in either direction, its cedar clinker-built sides and eighteen-foot oars were made to slip silently and swiftly through the water with its crew of six. 'Buoyant and graceful in her movements,' wrote Frederick Bennett, a British whaling surgeon, 'she leaps from billow to billow, and appears rather to dance over the sea than to plough its bosom with her keel.' At the rear, the mate gripped a great steering oar as he issued orders to men stripped for action; just as the boat's rowlocks were muffled, so they went shoeless so as not to scare their prey. A sperm whale was as ready to rear as a startled deer – and a gallied whale was good to no one.

Their pursuit was impelled by each boat's commander.

> 'Do for heaven's sake spring', the mate implored in whispered tones, 'The boat don't move. You're all asleep; see, see! There she lies; skote, skote! I love you, my dear fellows, yes, yes, I do; I'll do anything for you, I'll give you my heart's blood to drink; only take me up to this whale only this time, for this once, pull.'

They were the words of an urgent lover, just as the harpoons were the darts of a deadly Cupid; exhortations alternating between passionate blasphemies and competitive imprecations.

> 'Pull, pull, my fine hearts-alive; pull, my children; pull my little ones,' drawlingly and soothingly sighed Stubb to his crew . . . 'Why don't you break your backbones, my boys . . . Why don't you snap your oars, you rascals? . . . The devil fetch ye, ye ragamuffin rapscallions; ye are all asleep. Stop snoring, ye sleepers, and pull . . .'

So the small, lethal boats sped through the water, fast and fragile, ready if necessary to be turned into matchwood in the affray. As they drew near their prey, the oars were put aside as they waited.

And waited.

Sperm whales spend most of their time below the surface, and can sound for ten minutes or an hour. An experienced whaler knew how long an animal would stay down by its size: for every foot of whale they must wait a minute more.

It was a fearful calculation: the longer they waited, the greater the monster they faced.

A mile below, the whale might be scooping up squid in the silent depths, unaware of the danger that lurked above, the shapes that sculled over the ceiling of its world. But the time came when it needed to replenish the oxygen in its blood, returning to the light and air. The irony was that the sign of its renewed life – its characteristic angled blow, easily spotted from miles away – was also the signal for its demise.

Now came the moment for which these men had broken their backs. It too came shrouded in silence. 'Every breath was held; no one dared move a jot. The dropping of a pin in the boat might almost have been heard . . . Now we were within dart.' It was a meditation on what was to come, on the enormous task in hand. In this stillness was invested all the might of the whale versus the ingenuity of man.

They relied on the animal's design flaws: its blind spots, fore and aft. To approach a whale 'on the eye' was foolhardy; from its side it could see all that they were trying to do. So pulling head-on or behind, the boat crept as close as it dared. Through the surface they could see the fearsome flukes, three times the size of a man.

> How palpitating the hearts of the frightened oarsman at this interesting juncture! My young friends, just turn about and snatch a look at that whale. There he goes, surging through the brine which ripples about his vast head, as if it were the bow of a ship. Believe me, it's quite as terrible as going into battle, to a raw recruit.

This was the ultimate test, when each man would be judged; the moment on which their fortunes relied. It was also remarkably, almost stupidly dangerous: to pit a man against an animal so far in excess of him in size and power that even in the twentieth century, when hunting bottle-nosed whales – notorious for their ability to sound abruptly and take down a line with unbelievable speed – Norwegian ships would send out only single men, considering the task too hazardous for husbands with families.

Fear met fear. A harpooneer expected to spear a living creature one hundred times his size. A gigantic mammal startled by the appearance of an object it had never seen before. Through its very bones, connected to the auditory canal deep within its head, and through its startled eyes, protected by a film of oil, the whale sensed danger in unidentified noise and movement. Panic was its first response.

Once alerted, the entire school could swim off, at speed, invariably to windward. 'The slightest noise causes them to disappear with marvellous celerity,' as Charles Nordhoff observed. Giant whales could vanish into thin air. 'That's magic,' said Nordhoff's shipmate as one whale sounded with barely a toss of its head, so suddenly that 'it seemed just as though the vast mass had been suspended in space, and the suspensor had been suddenly cut asunder'. One minute a sixty-foot animal would be alongside them; the next, it had entirely vanished.

To gally a whale risked the failure of all that had brought the ship thousands of miles, captain and crew, provisions and whaleboats towards this one end. Sometimes the whale won even before battle was joined. Nelson Cole Haley's failure to harpoon a young, five-barrel calf as it dived after its mother ('I saw the shape of the little beggar under water', but his irons missed their target), earned him a volley of abuse and a confrontation with the captain back on board the *Morgan*.

More often than not the hunters were outwitted; proof, if it were needed, of the madness of whaling. Yet 'going on to a whale' was an intensely exciting moment; perhaps the most exciting thing these young men had ever done. It was 'glorious sport', rowing with their mates as they entered into the spirit of the chase, a rush of testosterone to coincide with a target on which to work out their rage. They were, in the argot of the time, bully boys, bully for the chase. This was why they forbore all the privations, for this one supreme moment, the adrenalin pumping in their arteries, even as the oxygen-rich blood coursed through the whale's.

Now the harpooneer rose to balance precariously at the prow, taking up his long iron from the crotch of the boat – the vessel and its weapons extensions of his power. As he stood, every muscle tensed towards the oncoming whale, the boat itself became a kind of brace, his right thigh set hard into a semicircle cut from the gunwale. This was the so-called clumsy cleat into which the hunter fitted, just as Ahab's peg-leg slotted into a socket made on the *Pequod*'s deck. Wood versus blubber; man's frail construction pitted against nature's formidable creation.

'Give it to him!'

Whaling was like war, 'actual warfare' in one whaler's eyes. For the young men in the boat, it was equivalent to going over the top; even more so for the man expected to throw the first blow for the first time. Only now did he realize the enormity of what he had to do, as he looked down into the water and the whale that seemed to fill his eyes. Some greenhands fainted at the sight, and had to be replaced by more experienced mates. Some went 'quite "batchy" with fright, requiring a not too gentle application of the tiller to their heads in order to keep them quiet'. Equally, the whale itself would react 'with affright, in which state they will often remain for a short period on the surface . . . lying as it were in a fainting condition', as if both man and whale were as shell-shocked as each other.

It was a military manœuvre, requiring superhuman strength. The harpooneer, rowing even harder than his mates, had at the last moment to drop his oar, pick up his weapon, and throw it twenty or thirty feet towards the whale; a man's straining blood vessels might burst with the effort, says Ishmael. At the crucial instant, the razor-sharp spear was released, hurtling through the air on its wooden stock, umbilically attached by the line as it whistled towards its target. More often than not it drew or failed to find its dreadful home. 'But what of that?' wrote Melville. 'We would have all the sport of chasing the monsters, with none of the detestable work which follows their capture.'

Time stopped still. Such was the intensity of the experience that, as their descendants would discover when rescuing rather than killing whales, the adrenalin of present danger obliterated all memory of anything else, even of the moment itself.

Harpooneer braced, power passing through iron to the whale.

Line curling in lazy loops, tightening to the fish.

Crew in mid-scull, every muscle tensed.

Mother ship on the horizon, fast fading into the distance.

Silence, before the clamour of life over death.

With a barely audible thud, the successful barb sank deep into blubber. With it all hell broke loose. The entire school of whales, feeling the blow communally, suddenly scattered to windward, causing the sea to erupt like an earthquake. Bucking and rearing, the harpooned whale tried to rid itself of the spear buried 'socket up' in its flesh. Sometimes the harpoon was bent double in the struggle. Its shaft was cast from flexible iron, so that it could be beaten back into shape, even if twisted to a corkscrew. As soldiers wore medals, so sailors kept such 'wildly elbowed' weapons as mementoes of their heroic encounters.

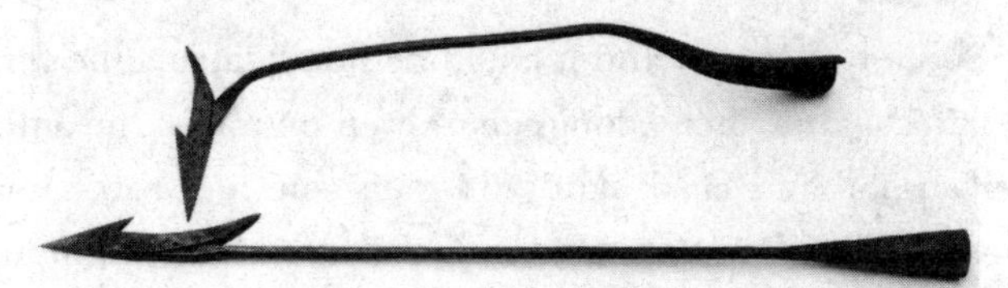

Now the whale would sound fast and deep, threatening to take its assailants with it. The line, long enough to run for a mile or more, paid out of its bucket where it lay like a coiled cobra, splashed with sea water to prevent it from burning with the friction and guided by hands covered with protective canvas 'nippers'. To sit with 'the magical, sometimes horrible whale-line', says Ishmael, was like sitting within a dangerous machine, 'the manifold whizzings of a steam-engine in full play, when every flying beam, and shaft, and wheel, is grazing you'. The whipping manilla rope could catch a man and yank him out of this world and into the next.

At one end, a sixty-ton animal. At the other, six men. Through the line they could *feel* the whale; an intimate connection between man and prey. The crew fought to haul the creature out of the depths as an angler tussles with a fish; an effort of resistance and power; a tug of war, or a tug of love. Suddenly, their enraged quarry surfaced with an almighty blow. Its very breath was fearful: sailors believed the spout to be acrid, able to burn skin or even, warns Ishmael, cause blindness, 'if the jet is fairly spouted into your eyes'.

Holding its buoyant, oil-filled head high out of the water, with its narrow jaw cutting the water below, the whale transformed itself 'from a bluff-bowed sluggish guillot into a sharp-pointed New York pilot-boat'. Now the terrified animal towed its tormentors on a Nantucket sleigh ride; at twenty-six miles an hour, this was the fastest any man had travelled on water: 'whole Atlantics and Pacifics seemed passed as they shot on their way'.

Sooner or later – and it could be hours later – the whale would tire. Only then, alongside or even on top of the animal itself, 'wood and black skin', did the scene reach its climax. Those rowing with their backs towards the whale might have been glad of their orders not to turn around. At any moment the whale might raise its tailstock twenty feet in the air, a towering slab of muscle so swift to deal death that it was called 'the hand of God'. With one flick it could send one of their number into eternity, an act as disdainful as theirs was arrogant. Worse still, the animal might actively turn on their craft, lunging with its toothed jaw held terrifyingly at right-angles to its body like a lethal saw. There was no defence against such an assault. It was man, or whale.

At the command, 'stern all', the harpooneer swapped place with the boat-steerer or mate, whose privilege it was, in that absolute hierarchy, to administer the *coup de grâce.* Drawing his long lance from its sheath, with both hands over the end to place his weight behind it, the mate plunged his iron in and out of the blubber. Blood running in rivulets over its black body, the maddened whale sought to wreak its revenge, impotently snapping its jaws open and shut. Then the blade found the life of the whale: the heart and lungs that lay behind its left flipper.

and they pierced his side with a lance

There it churned about like a poker until the cry went up, 'There's fire in the chimney!', its life-giving spout turned to a red fountain as thick blood pumped from the rapidly expanding and contracting blowhole. Now the whale entered its death flurry, swimming in a spiralling circle, the condemned animal vomiting up its final meal of squid, a pathetic reaction to its mortal internal wounds. With a juddering halt, its torment came to an end. 'His heart had burst!' And drawing its last breath, the whale rolled on its side, fin out, with one eye to the sky and – so its killers claimed – its head turned towards the sun.

they will look on the one whom they have pierced

For all the argot that served to distance them from their butchery, these were not men without hearts. They were not immune to the pathos of these scenes, to the death of something that represented life on such a scale. Charles Nordhoff would describe the wanton destruction he saw on his whaling cruise through the Indian Ocean and up the coast of Africa in search of sperm whales, as his crew mates harpooned and lanced any living thing they came across, from anaconda and hippopotamus to sea lion, as if anything alive became, by virtue of the fact, automatic targets. Young men like to kill things, sometimes just to see what happens.

And yet, when no sperm whales had appeared for weeks and the ship was driven to hunting humpbacks, even hard-bitten sailors objected to the killing of a mother and calf, the cow trying to protect her offspring by holding it tight to her body with her flipper or nudging it ahead and out of harm's way, only for the infant to fall prey to a well-directed lance. To one man, 'it was a useless waste of life . . . and besides had a tendency to excite the cow whale'. Later they saw one of the calves they had orphaned, now half-starved, desperately trying to suckle at a bull whale's belly, only to be violently driven off.

Men must eat, as must their families; their children must be shod, captains' houses must be shingled, their wives corseted; citizens must see by night. Their quarry was claimed with pennants, plaintively named 'waifs', planted directly into the whale's gaping blowhole. It was a final statement of possession: what was the whale's was now man's. These waifs also served to reunite the straying boats with their mother ship, perhaps miles away by now, perhaps even out of sight. Meanwhile, a sperm whale calf might nudge the whaleboat, searching the cedar sides for its mother's teats.

Physeter dolorosa

Rove through its flukes like a ring through a Moor's ear, the whale was chained and towed back, a fifty-ton dead weight dragged through the water at a mile an hour. If night had fallen by the time they returned, the whale would be secured to starboard, head astern. There it waited as the crew slept, their prey alongside, barnacle to barnacle, cosily safe until sunrise.

Then the real work began.

On the larboard or port side a section of the bulwark was removed, allowing a narrow cutting stage to be lowered, like a window-cleaner's platform, from which the mates, experts at the task, sliced at the whale with sharp spades. Other men dangled from ropes as whale mountaineers, hacking away to bring lumps of flesh and bone on deck, while their mates clambered over the slippery skin wearing crampon-spiked boots to carry out their delicate, brutal task. A hole was cut in the animal's side for the purchase of the giant blubber hook which swung from the mast. Thus the 'blanket' was unrolled, divesting the whale of what had given it warmth.

Pared off like the peel from a Christmas clementine, the result was cut into huge chunks and passed down to the blubber room. Here it was cut into manageable portions by half-naked men working in semi-darkness, often maimed by misaimed spades as the sharpened steel sliced off their own toes and fingers. Thick 'horse pieces' became 'bible leaves', thin slices to melt faster (while invoking images of the whale itself as a holy book). These were then hauled back up top and tipped into cast-iron try-pots set into brick ovens – strangely domestic structures, somewhere between blacksmith's furnace and kitchen range, as though someone had begun to build a house on deck.

For two days the work continued. Men laboured six hours on, six hours off, to the slithering, ripping, rippling, snapping sounds of torn tendons and sundered muscles, to the stink of blood and guts as the creature's severed head was separated into its constituent parts: the *case*, the chamber containing liquid spermaceti; the *junk*, the mass within the head; and the *white horse*, the fibres that held more oil in spongy cells. This was the rendering, a due process on this slavish ship, as the men in turn were

enslaved to the whale, paying obeisance to the vast creature dissected on deck: 'the entire ship seems great leviathan himself; while on all hands the din is deafening.' Most of the whale went to waste, chucked over the side to be gnawed by sharks and pecked by birds flocking to the scene.

As the animal came apart, it was in its blunt head that the hidden treasure was to be found: gallons of precious spermaceti. Ishmael takes us into this cavern , filled with a substance described by another as 'of a slightly rosy tint, looking like soft ice cream or white butter partly churned'. As man became part of the whale, the whale might even now take the life of a man. In a terrifying scene, Ishmael watches as Tashtego the harpooneer is lowered into the tun to bail out its spermaceti, only to fall in head-first, 'with a horrible oily gurgling'. The severed head bobs in the sea while the Indian struggles inside, about to drown in whale oil.

At that moment, a naked Queequeg appears, clutching a boarding-sword. Diving to the rescue, he pulls Tashtego out by his hair, delivering him from the fleshy pit like a Cæsearean-born baby, even as it threatens to become his grave. It would have been 'a very precious perishing', muses Ishmael, regaining his usual phlegm, 'smothered in the very whitest and daintiest of fragrant spermaceti; coffined, hearsed, and tombed in the secret inner chamber and sanctum sanctorum of the whale'.

> And the LORD appointed a great fish to swallow up Jonah; and Jonah was in the belly of the fish three days and three nights.

Such a deep-seated fear, of being engulfed by the whale, reached back to the Bible and beyond. The Victorian naturalist Francis Buckland described how one scientist had attempted a dissection of a beached sperm whale at Whitstable in 1829, descending

TRYING OUT.

into 'the gigantic mass of anatomical horrors', only to lose his footing and fall into the animal's heart, trapping his feet in its aorta. In the 1920s, an Oxford professor named Ambrose John Wilson sought to prove the possibility of Jonah's fate. He reasoned that only a sperm whale could have swallowed the prophet, baleen whales having throats that could admit nothing larger than a grapefruit. As it does not chew its food, the sperm whale uses strongly acidic stomach fluids to digest entire sharks and giant squid. 'Of course, the gastric juice would be extremely unpleasant but not deadly,' added the don, noting that the whale would digest only dead matter, lest it consume its own stomach.

In support of his theory, Wilson cited two case histories. In 1771 it was reported that a whaleboat working in the South Seas had been bitten in half by a sperm whale, and one of its crew seized by the assailant and taken down in its mouth as it sounded. Back at the surface, the animal disgorged the man, 'much bruised but not seriously injured', onto some wreckage. The historical distance made this story difficult to prove, but Wilson's second incident was recorded in 1891, when James Bartley of the *Star of the East*, then whaling off the Falklands, had disappeared into the water when a sperm whale's flukes lashed his boat. Hours later, the whale was killed and brought alongside the ship.

After working on the carcase all day and part of the night, the crew hauled its stomach onto the deck, and discovered their shipmate curled up inside, unconscious but alive. The man was laid out and given a sea-water bath to revive him; where he had been exposed to the animal's gastric juices, his skin had been bleached white, like some ghastly full-grown fœtus. For two weeks Bartley was a raving lunatic unhinged by his experience, only to recover his sanity and resume his duties. The captain's wife would later question the veracity of this story, but it encouraged those who believed a man could survive within a whale – although no one could explain how he could breathe in its belly.

More credible was another report by Egerton Y. Davis, a surgeon on the *Toulinguet*, sailing from Newfoundland in 1893 in search of harp seals, even if his account, too, is clouded by memory. As an old man, Davis recalled that one of the crew had slipped off an ice floe and into the jaws of an angered whale, which swallowed him before attacking the other sealers. Shot by the ship's cannon, the whale swam off in its death agony. It was recovered the next day, and when the crew cut into its gas-filled stomach, they found their mate.

It was a fearsome sight, said Davis, who proceeded to deliver a pathological description. The young man's chest had been crushed by the animal's jaws, so he was probably already dead by the time his body reached the whale's stomach. Gastric mucosa covered the victim like the slime of a giant snail; it was particularly thick on those parts of his flesh that were exposed: his face, his hands and part of his leg where his trousers were torn; these areas were macerated and partly digested. Oddly enough, the lice on his head had survived.

The surgeon sought to reassure his shipmates that the man had not suffered. 'It was my opinion that he had no consciousness of what happened to him.' The idea that the victim might have been aware as he was swallowed was too terrible to contemplate; although in secret his fellow sailors may have wondered what it was like to be within the belly of the whale, to slither down its gullet like a whiting down a gannet's neck and into the nameless horror of the leviathan's maw.

Such stories would persist, from the whale that gulped down Pinocchio, to George Orwell's *Coming up for Air*, in which the narrator recalls his Edwardian father reading of 'the chap . . . who was swallowed by a whale in the Red Sea and taken out three days later, alive but bleached white by the whale's gastric juice', adding that 'he turns up in the Sunday papers about once in three years'. Indeed, in a letter to *The Times* in 1928, a correspondent claimed to have met a missionary to the Southern Whaling Fleet who was swallowed by a sperm whale. For a man of the cloth, he appears to have been rather accident-prone, having often fallen overboard – a regular Jonah – but 'could hold his breath longer than most men'. More fortuitously, his shipmates had seen him fall, and harpooned the whale which, in its flurry, evacuated its stomach, and the indigestible cleric along with it.

> And the LORD spoke to the fish, and it vomited out Jonah upon the dry land.

Evidently fascinated with such stories, Orwell elaborated on the theme in a famous literary essay written just as the Second World War broke out. *Inside the Whale* saw something strangely appealing in the idea:

> the fact is that being inside a whale is a very comfortable, cosy, homelike thought . . . The whale's belly is simply a womb big enough for an adult. There you are, in the dark, cushioned space that exactly fits you, with yards of blubber between yourself and reality . . . Even the whale's own movements would probably be imperceptible to you. He might be wallowing among the surface waves or shooting down into the blackness of the

> middle seas (a mile deep, according to Herman Melville), but you would never notice the difference. Short of being dead, it is the final, unsurpassable stage of irresponsibility.

Allegory or tall tale, such notions merely lend more mystery to the whale; an animal so strange and savage and innocent, so monumental in man's imagination now reduced to bits on the deck of a ship.

So the process continued. The jaw was wrenched from its cartilaginous hinges, the conical teeth yanked out as if by some cetacean dentist. One whale could yield forty or fifty fist-sized pieces of sea-ivory, issued to sailors for scrimshanding, work for idle days when whales were few. Some teeth might be swapped for supplies; they were highly valued in Fiji, where the captain of the *Morgan* exchanged sperm whale teeth for food far in excess of their value on the streets of New Bedford, where, as young Haley noted, they'd fetch a dollar fifty at the most.

By now the deck was awash with oil, one great slick sliding rink; men might slip off and into shark-infested waters. Life was tentative: others could be crushed by lumps of whale, or splashed with boiling oil, or sliced by flenshing knives. Compared to such perilous butchery, the sorting of spermaceti was a popular chore. Collected into tubs, sailors squeezed the lumps from the oil which coagulated as it cooled away from the heat of the body. Some climbed into the tubs themselves like grape-tramplers, pulling out the fibrous integuments which would mar the superior quality of the product.

'No king of earth, even Solomon in all his glory, could command such a bath,' wrote one whaler. 'I almost fell in love with the touch of my own poor legs, as I stroked the precious ointment from the skin.' The task imparted a feminine air to otherwise grisly and dangerous duties; for the narrator of *Moby-Dick*, it induced an erotic reverie as his fingers began to 'serpentine and spiralize' like eels and he was lulled by the scent and sensuality. In the easily stirred Ishmael, such 'sweet and unctuous duty' becomes a kind of Blakean transcendence, and 'in thoughts of the visions of the night', he sees 'long rows of angels in paradise, each with his hands in a jar of spermaceti'.

Elsewhere, a hellish scene held sway. As the try-pots were heated, the flames were fed with slivers of blubber called 'cracklings'; thus the whale cooked itself. Naturally, such an irony did not escape Ishmael. 'Like a plethoric burning martyr, or self-consuming misanthrope, once ignited, the whale supplies its own fuel and burns by his own body.' And as darkness fell, the flickering red light turned it all into an infernal vision akin to Loutherbourg's painting of the ironworks at Coalbrookdale, satanic womb of the Industrial Revolution; or something more apocalyptic:

> the wild ocean darkness was intense. But that darkness was licked up by the fierce flames, which at intervals forked forth from the sooty flues, and illuminated every lofty rope in the rigging, as with the famed Greek fire. The burning ship drove on, as if remorselessly commissioned to some vengeful deed.

Notions of horror mar these honest acts of industry in our eyes. What did Melville feel at the time, as he watched, and took part in such scenes conducted far from civilized gaze? Words had the power to conqueror memory; but they were useless in the catching and rendering of whales, save to supply captions to Victorian engravings: '*There she blows!*', '*Whereaway?*', '*She has fire in the chimney!*'

After it was all done, the ship was scrubbed; in another example of cetacean self-sufficiency, unrefined sperm oil possessed 'a singularly cleaning virtue', and 'the decks never look so white as just after what they call an affair of oil'. But no sooner was the place clean and its crew with it, 'the poor fellows just buttoning the necks of their clean frocks', than the lookouts would shout,

There she blows!

and they would 'fly away to fight another whale, and go through the whole weary thing again'.

Ah the world. Oh the world.

VII

The Divine Magnet

To produce a mighty book, you must choose a mighty theme.

The Fossil Whale, *Moby-Dick*

Having been halfway round the world, Melville returned to his family in sleepy Lansingburgh in October 1844. He was only twenty-five, yet he had seen more in three years than most people would in a lifetime. He had been away for so long and so far from home that he'd almost forgotten who he was, or who he was supposed to be: hero, or outcast? Encouraged by his sisters, he wrote down the stories he told them of his adventures in the South Seas where, with his 'remarkably prepossessing' friend Toby Greene, a black-eyed, curly-haired boy of seventeen, he had deserted the *Acushnet* and lived among naked savages.

Typee – the word means man-eater, although Melville feared having his face tattooed with the devil's blue more than being consumed by his hosts – was a sensation among the men of an American renaissance keen to distinguish itself from British literature. It was a sensual, sometimes idyllic account of life among the natives of the Marquesas Islands, as well as being a critique of the western influences beginning to taint their paradise. Walt Whitman saw it as a 'strange, graceful, most readable book . . . to hold in one's hand and pore dreamily over of a summer's day', while Nathaniel Hawthorne admired its 'freedom of view' and tolerance of 'morals that may be little in accordance with our own; a spirit proper enough to a young and adventurous sailor'. It turned Melville into America's first literary sex symbol – an almost disreputable figure.

A year later, as if licensed by his literary success, Melville married Elizabeth Shaw, daughter of his father's friend, Lemuel Shaw, a wealthy Boston judge. The couple settled at 103 Fourth Avenue, New York, where Melville became part of the circle known as Young America which revolved around the editor Evert Duyckinck and his house on Clinton Street. But the sequels he wrote to *Typee* – *Omoo, Mardi* and *Redburn* – did not fare as well, being judged degraded, immoral, even grotesque, and late in 1849 Melville left his young wife and baby son, Malcolm, for England, where he hoped to sell his latest book, *White-Jacket,* and, perhaps, finance further travels. He sailed from a wet and rainy New York that October on the liner *Southampton,* and two weeks later arrived at Deal, from where he made his way to London and a fourth-floor room off the Strand, 'at a guinea & a half per week. Very cheap.'

Not many people walk down Craven Street now, even though it lies off one of London's busiest thoroughfares. Hidden behind Charing Cross Station, its blackened brick Georgian houses seem remaindered from the modern city. Number 25 is at the end of the terrace, with a wide bow window at the side. At the top of its winding, uneven staircase are attic rooms, usually the preserve of servants. Their view is restricted now, but before the Thames was embanked and houses still ran down to the river, Melville could look out of his room onto an imperial waterway coursed by boats and barges.

London was rising in a slew of stone and brick, of movement and noise. Nearby were the recently built Trafalgar Square and the National Gallery, while the new Palace of Westminster, still under construction, loomed over the water; the sun seldom shone on its intricate façade, obscured as it was by the fog that both cloaked and sustained the city. Stepping out from his boarding house and into the Strand, the American wore a new green coat, the source of 'mysterious hints dropped' on board the *Southampton.* He looked recognizably other, a Yankee in the court of Queen Victoria.

In his travel journal, one of the few documents that details his life, Melville recorded the 'dark & cozy' inns of the City, the Cock Tavern, the Mitre, the Blue Posts, and the Edinburgh Castle, where he drank Scotch ale and ate chops and pancakes – Herman had bad manners, and often spoke with his mouth full – talking metaphysics with Adler, a German scholar whom he had met on the voyage over. He saw the sights, visited the galleries and even attended a public execution; Dickens was among the same crowd. He also touted *White-Jacket* round the publishers, with little success. But as he roamed London, other ideas were forming in his head.

'Vagabonding' through alleys and 'anti-lanes' from the new Blackwall Tunnel to Greenwich and back to Tower Hill, Melville passed the place where a well-known beggar, a former sailor with one leg, wore a painted board displaying the circumstances of his loss. The scene was an echo of the unfortunates in Liverpool, only here the picture was more terrifying: 'There are three whales and three boats; and one of the boats (presumed to contain the missing leg in all its original integrity) is being crunched by the jaws of the foremost whale'. London itself was a whaling port. The south-eastern docks at Rotherhithe were home to whale-ships and processing plants, while famous entrepreneurs of the trade ran their businesses from the more genteel address of the nearby Elephant and Castle.

Whales were on Melville's mind; sometimes it seemed they were swimming down the city's streets. The visceral butchery of Fleet Market reminded him of a blubber room; returning home at two in the morning from a 'snug' evening with some young Londoners, he 'turned flukes' in Oxford Street. It was as if the imperial metropolis were rousing the spirit of Moby Dick. In his attic room, high above the gas lamps shining on the midnight streets, Melville mourned his elder brother, who had worked and died in this city. 'No doubt, two years ago, or three, Gansevoort was writing here in London, about the same hour as this – alone in his chamber, in profound silence . . .' That night he was plagued by 'one continuous nightmare till daylight'. He blamed it on strong coffee and tea; but perhaps whalish monsters were stirring in his dreams.

After a brief trip to the Continent – his intention to travel to the Holy Land circumscribed by London's refusal to pay more than a reduced sum for his book – a homesick Melville sailed from Portsmouth to New York, where he set to work on a new novel – an unashamedly commercial venture. It was to be 'a romance of adventure, founded upon certain wild legends in the Southern Sperm Whale Fisheries', he told his English publisher, Richard Bentley. In what may have been an almost desperate move, Melville turned to his whaling experiences to capitalize on a new commercial empire at home, one that combined the American talents for heroism and consumerism.

New York was more prosperous and bustling than ever, a rival to London's imperial sway. The profits from whales funnelled through this city, too. It was a place of import and export, its masts and piers reaching out to other lands, even as it sent its equal sons and equal daughters around the world. Close to Wall Street, where his brothers worked, was Nassau Street and its publishing and newspaper offices, Manhattan's equivalent of Fleet Street and the Strand. Nearby were the luxurious new Astor House Hotel, and the Shakespeare Tavern where writers such as Washington Irving and Edgar Allan Poe drank. Around the corner was Barnum's American Museum which, that summer, was decorated with a huge canvas banner advertising the whale that lay within.

As much as *Moby-Dick* was a product of Melville's adventures at sea, it was also born of the city; its opening scenes state as much, set as they are on the quayside at the end of Pearl Street. In a strange, allusive manner, New York itself became the White Whale, just as Joseph Conrad would see Brussels as a whited sepulchre built on human bones, and as Gansevoort Melville had seen London as the modern Babylon. Even the island of Manhattan was whale-shaped, a pallid behemoth, both fascinating and appalling. Here, on what purported to be dry land, Melville's desires were ambivalent. Expounded in his book, they represented both liberation and dread, deep longings and profound fears. And symbolic above them all was the whale: the leviathan that had risen from the deep to take hold of his imaginings.

In his years at sea, Melville heard tales of lethal encounters between man and whale. Now, as Yankee whaling reached its peak, these incidents seemed to be becoming ominously more frequent. The whales were fighting back, breaking bones and boats, drowning men, turning on their assailants with a vengeful intelligence. On 15 August 1841, for instance, soon after the *Acushnet* left port, another New Bedford vessel, the *Coral*, encountered a school of sperm whales one hundred miles south of the Galápagos. The captain, James H. Sherman, recorded that having struck one whale, the beast rounded on the whaleboat that pursued it, 'and chewed her in many Hundred Pieces'.

'Spouting good blood while Eating the Boats', the animal set off, followed by its hunters, but as they drew close and the mate was about to lance it, the whale 'turned upon him and Eat his Boat up also'. In the chaos, the captain dived in to save a drowning crewman, and brought him back to the boat; but the whale had not finished with them. In its flurry, it turned on its side, its jaw lunging at the captain. Only then did Sherman manage to 'hove an Iron into him ... and in a few moments he was in the Agonies of Death and Breathed his Last'.

As he began to research his story, Melville found other accounts of avenging whales. The *Union*, a Nantucket ship, was lost off the Azores in 1807 after an attack by a whale, while a Russian vessel was raised three feet out of the water by 'an uncommon large whale . . . larger than the ship itself'. Nor were sperm whales the only cetaceans able to stove a ship. Grey whales were called devil fish for their propensity to turn on their hunters, and fin whales, too, were known to charge and sink a vessel. Even smaller whales could be dangerous: at least one sailor was killed by a blackfish during Melville's years of whaling.

But it was the otherwise placid sperm whale that could do the most damage. In 1834 Ralph Waldo Emerson was riding in a stagecoach when he had heard a sailor talk of a white whale called Old Tom which attacked with its jaw, '& crushed the boats to small chips . . . A vessel was fitted out at New Bedford, he said, to take him.' Gathering up these stories, Ishmael speaks of a confederacy of dæmonic whales which gained 'an ocean-wide renown', a veritable champions' league: Timor Jack, 'scarred like an iceberg', a fearsome fighter who was only caught when a barrel lashed to the end of a harpoon with which he was tapped on the shoulder, distracted his attention while 'means were found of giving him his death wound'; New Zealand Tom, which destroyed nine boats before breakfast and was 'terror of all cruisers . . . in the vicinity of the Tattoo Land!'; and Don Miguel, another grizzled battler, 'marked like an old tortoise with mystic hieroglyphs upon the back!'

Of all such whales, the most vivid – because it came as a first-hand testimony – and the most infamous was the one that sank the *Essex*, an account of which was published in 1821 by the ship's mate, Owen Chase. His title summed up the story sensationally, if not succinctly:

> NARRATIVE OF THE MOST EXTRAORDINARY AND DISTRESSING SHIPWRECK OF THE WHALE-SHIP ESSEX, OF NANTUCKET: WHICH WAS ATTACKED AND FINALLY DESTROYED BY A LARGE SPERMACETI-WHALE IN THE PACIFIC OCEAN.

In his book (which to Melville bore 'obvious tokens' of having been dictated), the aptly named Chase describes how a bull sperm whale, apparently enraged by attacks on his fellow whales, came at the *Essex* at 'twice his ordinary speed', with 'tenfold fury' and 'vengeance in his aspect', his tail thrashing and his head halfway out of the water – a truly terrifying sight. Hitting the ship full-on, the whale smashed into her bows, then swam off to leeward and was not seen again. The resultant exchange between Captain Pollard and his first mate might have come from a 1940s British film.

> 'My God, Mr Chase, what is the matter?'
> 'We have been stove by a whale.'

As the *Essex* sank, her crew were circled by the animals they had hunted, the whales unseen in the darkness, 'blowing and spouting at a terrible rate'. Drifting on the ocean in open boats, the shipwrecked men could hear huge flukes thrashing furiously in the water, 'and our weak minds pictured out their appalling and hideous aspects'. Yet it was not the whales they had to fear: it was their fellow man. The starving and thirst-maddened survivors refused to sail towards nearby islands for fear of their cannibal inhabitants – only to end up eating each other to stay alive.

Melville claimed not only to have met Chase's son, who lent him a copy of his father's book – 'The reading of this wondrous story upon the landless sea, & close to the very latitude of the shipwreck had a surprising effect upon me' – he also maintained he had seen Owen Chase himself on his ship, the *William Wirt*. However, by the time Melville was sailing on the *Acushnet*, Chase had retired from the sea and was living alone in Nantucket, hoarding food in his attic, still fearing starvation and having lost his mind, clutching his friend's hand as he sobbed, 'Oh my head, my head'. Meanwhile, close by, his former captain lived with his own awful memories. Distrusted with any new command, Pollard worked as a nightwatchman and lamplighter, wandering the streets of Nantucket as if to atone for his sins. It was only after he wrote his book, on his first visit to an island that he had only imagined until then, that Melville met 'Capt. Pollard . . . and exchanged some words with him. To the islanders he was a nobody – to me, the most impressive man, tho' wholly unassuming, even humble, that I ever encountered.'

As Melville's imagination fastened on the story of the *Essex*, it was supplemented by other legendary whales in print. In 1839 Jeremiah Reynolds's 'Mocha Dick: or, the White Whale of the Pacific' was published in the *Knickerbocker Magazine*. A friend of Edgar Allan Poe's, Reynolds was an eccentric writer and explorer who believed in a hollow earth. He embroidered on tales of a white whale known to haunt the waters off the Chilean island of Mocha, 'an old bull whale, of prodigious size and strength. From the effect of age, or more probably from a freak of nature, as exhibited in the case of the Ethiopian Albino, a singular consequence had resulted – *he was as white as wool!*'

This eerie creature was claimed to be one hundred feet long, rugged with barnacles and able to shatter boats with his twenty-eight-foot-wide flukes, or grind them to pieces with his massive jaws. He was said to have killed thirty men, stoven fourteen boats, and had nineteen harpoons planted in him. Reynolds's story ends with the whalers triumphant: 'a stream of black, clotted gore rose in a thick spout above the expiring brute, and fell in a shower around, bedewing, or rather drenching us, with a spray of blood . . . And the monster, under the convulsive influence of his final paroxysm, flung his huge tail into the air . . . then turned slowly and heavily on his side and lay a dead mass upon the sea.' In reality, Mocha Dick – or at least a whale like him – continued to roam the oceans from the Falkland Islands to the Sea of Japan, attacking English, American and Russian ships without discrimination before being taken by a Swedish whaler in August 1859.

It was as if the hunted whale had become aware of its persecution, and was fighting a rearguard action. 'From the accounts of those who were in the early stages of the fishery,' wrote Owen Chase, 'it would appear that the whales have been driven, like the beasts of the forest, before the march of civilisation into remote and more unfrequented seas.' 'Sperm whales are now much scarcer than in years past,' noted Charles Nordhoff in the 1850s, 'owing to the number of vessels which annually fit out from American and various parts of Europe, partly or entirely in pursuit of them.'

They may also have been more formidable opponents. Chase claimed that the animal that sank the *Essex* in its 'mysterious and mortal attack' was eighty-five feet long; Thomas Beale recorded sperm whales of eighty feet; while a lower jaw preserved in Oxford's University Museum confidently announces an owner of eighty-eight feet. In *Nimrod of the Sea; or, The American Whaleman*, published in 1879, W.M. Davis registered sperm whales reliably measured at ninety feet; Ishmael heard of others one hundred feet long. Yet no modern sperm whale grows to more than sixty-five feet.

Some speculate that the hunting of large whales has gradually reduced their genetic likelihood; perhaps the *Essex*'s assailant was the last of a gigantic breed. The larger lone bulls were inevitably the first to be taken, and twentieth-century hunting accelerated this cull, while skewing our knowledge of the whale's longevity. Assessments of their life spans rely on whaling statistics from the second half of the last century, by which time most of the older animals – being larger and more profitable – were dead.

By the end of worldwide whaling, nearly three-quarters of all sperm whales had been killed, reducing a population of more than a million in 1712 to 360,000 by the end of the twentieth century. Even in the 1840s the whalers saw a definite decline, and wondered if their efforts would lead to the animal's demise. In the chapter entitled, 'Does The Whale's Magnitude Diminish? – Will He Perish?', the impeccably informed Ishmael cites the buffalo as 'an irresistible argument . . . to show that the hunted whale cannot now escape speedy extinction', although he also declares that sperm whales that once swam as 'scattered solitaries . . . are now aggregated into vast but widely separated, unfrequent armies'.

Were these animals collectively enraged by their attackers and determined to fight back, just as modern rogue elephants, their habitat destroyed by man, are now thought to turn on humans? If the scars on bull male sperm whales are any indication, they are ferocious fighters among themselves. Certainly, the Yankee captains thought that the whales had become more wilful. Docile beasts turned on their assailants, using their own weapons – jaws, heads, flukes. Captain Edward Gardner of the *Winslow*, out of New Bedford, was another victim, nearly killed by a sperm whale off Peru in 1816, 'wounding me on my head' and 'breaking my right arm, and left hand badly lacerated, my jaw and five teeth were broken, my wounds bled copiously'.

It was as if the whales were complicit in the role allotted to them. 'In times past, when they were not so continually worried and followed, they were much easier to approach, although often giving battle when attacked,' Charles Nordhoff observed. 'Now, however, the utmost care is required to "get on".' As Ishmael confides, 'I tell you, the sperm whale will stand no nonsense.'

And yet, conversely, the whales' reactions could be entirely and almost pathetically inactive. Although a sperm whale could easily outdistance its persecutors, diving far and fast out of range, it often did not do so. Sometimes when their enemies approached, or when one of their number was injured – as Frederick Bennett wrote in another of the books that Melville consulted during his researches – the whales would 'crowd together, stationary and trembling, or make but confused and irresolute attempts to escape'.

Paradoxically, such suicidal behaviour was in part due to the animal's ability to live in the depths. At the surface, the sperm whale is slower, less agile and has less time and energy than other whales – and is therefore less able to flee such an unnatural predator as man. It is an inexplicable and potentially fatal evolutionary flaw, and it led the writer John Fowles to wonder why the sperm whale 'has never acquired – as it easily could in physical terms – an efficient flight behaviour when faced with man. At times, it will almost queue up to be gunned . . . The poor brutes just never learnt.'

Man, whale, life, death: this was the story Melville had to tell. No writer, before or since, could have had such an epic gift. On one side, the world's greatest predator, more legendary than real; on the other, young American heroes, men who risked everything in the pursuit of oil. Theirs was a quest that asserted the myth of America, the great new democracy in which anyone might find their fortune; but it also brought them into contact with something more mysterious. Moby Dick was a spectral creature believed to be omnipresent – 'actually . . . encountered in opposite latitudes at one and the same instant of time' – and able to escape repeated and bloody attacks, reappearing 'in unensanguined billows hundreds of leagues away'. In this incarnation, the whale became ubiquitous, its hugeness as numinous as dark matter; an animal more mystical than muscular; as if the spermatozoid were a universe at the same time.

At first Melville dismissed such metaphysics. His book was to be as much a commercial venture as any whale-ship setting sail from New Bedford, his lay to be shared with his publishers. 'Blubber is blubber,' he told a friend, treating his new work as another *Redburn*, which he knew 'to be trash, & wrote it to buy some tobacco with'. But all that would change. In his magpie imagination, named and nameless terrors gathered strength and power like the ominous white whale seen below the surface, 'with wonderful celerity uprising, and magnifying as it rose . . . his vast, shadowed bulk still half blending with the blue of the sea'. In the process *Moby-Dick* became a legend itself; a story encoded with its own terrible beauty, one that saw into the future even as it looked into the past.

* * *

Monument Mountain stands off Route 7, its lower reaches surrounded by dense woods. A century and a half ago, the trees were not so close-grown. On a summer morning, the aftermath of two days' rain is still percolating through the pines, drip-drip-dripping as I clamber my way up the slippery path. The hillside is strewn with huge boulders; a deep valley opens to the other side, coursed by a stream overhung by ferns. As I make my final ascent, a rain cloud bursts overhead, sweeping over the rocks on which garter snakes bask; bright orange lizards dart into crevices. At the summit, quartzite crags topped with stunted pines hang precipitously over themselves. Far below is the green-carpeted valley of the Housatonic River. Hawks hover on the updraught. All the world seems caught in the stillness.

It was here in Western Massachusetts, in the summer of 1850, away from 'the heat and dust of the babylonish brick-kiln of New York', that Melville met a man who would change the course of his life. While staying with his aunt in nearby Pittsfield, he read Nathaniel Hawthorne's *Mosses from an Old Manse,* and was besotted with its wistful evocation of old New England. By coincidence Hawthorne himself was living nearby, drawn to the sublime beauty of the Berkshires – countryside not unlike England's Lake District. It was a romantic setting in the purest sense of the word; and what happened next was a kind of epiphany.

At forty-six years old, Nathaniel Hawthorne was America's most famous writer. He too came from a sea-going family – he was only four when his father, a sea captain, had died of fever in Surinam – and he had grown up with his mother and two sisters. That much he and Melville had in common. But where the sea was Herman's Harvard and Yale, Nathaniel had attended the grassy campus of Bowdoin College, Maine, before exchanging it for a gloomy house in Salem, where he spent twelve years sequestered in his attic, emerging only at night to walk the streets. 'I have made a captive of myself, and put me into a dungeon,' he confessed; 'and now, I cannot find the key to let myself out.'

'Handsomer than Lord Byron', with dark eyes 'like mountain lakes seeming to reflect the heavens', Hawthorne dwelt on morbid things, although the monsters he summoned were decidedly human. His Puritan ancestors – with 'all the Puritanic traits, both good and evil' – had persecuted Quakers, and had taken part in the Salem witch trials. This legacy infused the fictional world Hawthorne inhabited, and the real world he invented. He was, as the poet Mary Oliver would write, 'one of the great imaginers of evil'.

Hawthorne was filled with regret at the way the world had been, and the way it was becoming. 'Here and there and all around us,' he wrote in his story, 'Fire Worship', '. . . the inventions of mankind are fast blotting the picturesque, the poetic, and the beautiful out of human life.' He once told his wife Sophia that he felt as if he were 'already in the grave, with only life enough to be chilled and benumbed'. And although he loved to swim in the river at the bottom of his garden in Concord at night, seeing the moon dance on the surface – where I swam too, pushing my way through the clear water and bright green weeds, imagining Billy Budd caught in their oozy fronds – he was haunted by the memory of a drowned young woman who was once pulled from the same river, her limbs white and swaying in the water.

Hawthorne was, in his own words, 'a man not estranged from human life, yet enveloped in the midst of it, with a veil woven of intermingled gloom and brightness'. He wrote artful allegories burdened with the weight of history, guilt and revenge; especially the stories that Melville saw as Hawthorne's masterpieces, and which would influence his own work. In 'Young Goodman Brown', set in seventeenth-century Salem, a young man is summoned to the forest at night to find the entire town enslaved to the devil, even his young wife. In the futuristic 'Earth's Holocaust', a bonfire on the prairie incinerates every example of human excess, from tobacco to works of literature. Yet one thing will not burn in this reforming pyre: the latent evil in every human heart. Sin, too, was the subject of his novel, *The Scarlet Letter,* published in 1850; and in the wake of its success, Hawthorne had escaped the clamour of fame by moving to Lenox in the Berkshires, close by a calm freshwater lake, where he hoped to work on his next book, *The House of the Seven Gables.*

Hawthorne could not avoid society even in the country, and on 5 August he was persuaded to attend a picnic organized by David Dudley Field, a well-connected New York lawyer. The guests included distinguished literary figures: Evert Duyckinck, Oliver Wendell Holmes – coiner of the term, Boston Brahmin – 'several ladies' and Melville. The party set off for Monument Mountain, but before they could reach the summit a sudden shower sent them running for shelter under a rocky ledge, where they drank champagne from a silver mug.

As the sun reappeared, the picnickers struck out for the mountain top. Melville was in high spirits; perhaps the alcohol and the rarefied air had gone to his head. He clambered over a long rock which jutted out like a bowsprit, pretending to haul in imaginary rigging, and made as if to harpoon a whale-shaped pond in the valley below. The young man's play-acting was a burst of energy in the dog-days of summer – an echo of the scenes in *Typee* in which the narrator and his fellow deserter Toby climb a tropical peak to escape the tyranny of their ship, and feel the intensity of their new-found freedom.

The headiness of the day, the sublimity of the landscape, and, perhaps, Melville's company, were infectious, and they roused Hawthorne to similar antics. That afternoon, as they wandered through the 'Gothic shades' of a gloomy spot known as the Icy Glen – it was said ice was found in its mossy recesses all year round – it was his turn to perform, shouting out, in his rich voice, 'warnings of inevitable destruction to the whole party'. Then they all repaired to the Fields' house for dinner, at which they discussed the sea serpent that had made an appearance off the coast of Massachusetts.

It was clear that Hawthorne – already an admirer of *Typee* – found Melville a magnetic figure. 'I do not know a more independent personage,' he would write. 'He learned his travelling-habits by drifting about, all over the South Sea, with no other clothes or equipage than a red flannel shirt and a pair of duck trousers.' Perhaps he even listened with envy to the sailor's adventures, a sense of outlandish experience to contrast with his own haunted introspection. That day on the mountain marked an almost alchemical mix: of fire – Hawthorne's prairie holocaust – and water – Melville's whalish romance. Both were men of a brave new republic; both might have looked optimistically towards the future. But in time, the lively and mercurial Melville would descend into the gloom that Hawthorne inhabited, swapping the sun-baked summit for the dank dripping glen.

A month after meeting Hawthorne, Melville moved to a farm two miles south of Pittsfield, bought with the help of his wealthy father-in-law and named Arrowhead after the Indian artefacts he found in its fields; in the distance stood Mount Greylock, the highest point in Massachusetts. For two hours a day Melville would work the fields as a farmer; he even sold cider from the roadside, a memory of the house's former guise as a tavern. But he was also less than an hour's ride from Hawthorne's house at Lenox. 'I met Melville, the other day,' Hawthorne told a friend, 'and I like him so much that I have asked him to spend a few days with me.'

Melville expressed himself in rather stronger terms. In a gesture that was both revealing and concealing at the same time, he wrote a review of *Mosses from an Old Manse* in the guise of 'a Virginian spending July in Vermont', and in language that seems astonishingly suggestive to modern ears: 'I feel that this Hawthorne has dropped germinous seeds into my soul. He expands and deepens down, the more I contemplate him; and further and further, shoots his strong New England roots in the hot soil of my Southern soul.'

Tethered by meetings and ever longer letters, the friend-ship between the two men grew. Later, when Sophia Hawthorne and their daughters, Una and Rose, went to visit relatives, leaving Nathaniel in charge of five-year-old Julian and his pet rabbit, Melville took the opportunity to call. He arrived, glamorously, driving a barouche and pair, with Evert and George Duyckinck, his dog and a picnic in the back. Hawthorne supplied the champagne, and they set off to visit the Shaker village at Hancock. Hawthorne, who had sampled Utopia during his brief stay at the Transcendentalist commune, Brook Farm, found the celibate Shakers a sad blasphemy, with their 'particularly narrow beds, hardly wide enough for one sleeper, but in each of which, the old elder told us, two people slept'. It was a 'close junction of man with man' which Hawthorne professed to find 'hateful and disgusting'. Ishmael and Queequeg may have seen it differently.

At Lenox, the two men would sit in the Hawthornes' parlour smoking cigars normally forbidden in the house, talking 'about time and eternity, things of this world and of the next . . . and all possible and impossible matters, *that lasted pretty deep into the night*' (a phrase that Sophia inked out when editing her husband's journal for publication). They did not agree on all subjects: on slavery, for instance, for whose victims Hawthorne had not 'the slightest sympathy . . . or, at least, not half as much for the labouring whites, who, I believe, as a general thing, are ten times worse off'. For all Melville felt for Hawthorne, it seemed he wanted more than his friend could give.

Melville's book – which he described to Evert Duyckinck as 'a romantic, fanciful & literal & most enjoyable presentment of the Whale Fishery' – was almost finished when he came to the Berkshires. Meeting Hawthorne changed all that. The younger man had complained of being restrained from writing 'the kind of book

I would wish to'. Now he was compelled to see the significance of his experiences, and as if to set them in context, he began to read rapaciously, as though he had never read before: books brought back from London, such as Mary Shelley's *Frankenstein*, or ones borrowed from the New York library, such as William Scoresby's *An Account of the Arctic Regions*; Robert Burton's eccentric and digressive *Anatomy of Melancholy*; essays by Emerson in which God was revealed in nature; and Thomas Carlyle's *Sartor Resartus*, imbued with dreams, dæmonic possession and self-sacrificing love.

Then he found a complete edition of Shakespeare's plays in print large enough to overcome his weak eyes. 'I would to God Shakespeare had lived later, & promenaded in Broadway,' he fantasized. But he would also fill his own book with earthy asides and euphemisms; jokes about chowder and bar-room quips with which Ishmael wryly undermines his creator's high-flown words, declaring at one point that he regards the whole dangerous voyage of the *Pequod* – and life itself – as a 'vast practical joke', and informs Queequeg that he 'might as well go below and make a rough draft of my will', with his friend as his lawyer, executor, and legatee.

Melville was liberated by America, a place where he could write about anything and everything, and where he was perfectly aware of the double meaning of his words, even as Starbuck exhorted his crew: 'Pull my boys! Sperm, sperm's the play!' There was a new urgency to his work which almost seemed to set him apart from what he was doing, time-coding his words –

> . . . that down to this blessed minute (fifteen and a quarter minutes past one o'clock P.M. of this sixteenth day of December, A.D. 1850), it should still remain a problem, whether these spoutings are, after all, really water, or nothing vapour . . .

– as if he were suddenly able to see beyond himself and into the whale, in an out-of-body experience even as he moved towards it. Like Ishmael, he felt reborn. 'Until I was twenty-five I had no development at all,' he told Hawthorne. 'From my twenty-fifth year I date my life.' Something fused into one headlong effort, as great as his quarry, as great as the industry it commemorated. With sprawling ambition and an utter lack of convention, Melville crossed latitudes of time and space, blurring them as he did so, constantly reiterating, 'All this is not without its meaning', laying meaning upon meaning, drawing himself on, writing and re-writing obsessively, creating a kind of exclusion zone to which his own wife Lizzie could only gain admittance by knocking incessantly on his door until he deigned to answer.

He had recreated the conditions on board ship inside his study and in his mind, and in the process *Moby-Dick* changed from a romance to a fearful, fated work. Parts of the book seem to be written automatically, as if possessed by the spirit of the White Whale, the Shaker God incarnate. There was something forbidden about his subject, named for a mythic Mocha Dick but which also elided with the name of his fellow deserter, the dark and prepossessing friend whom he had thought dead but had met again in Rochester, New York. 'I have seen Toby, have his darguerrotype [*sic*] – a lock of the ebon curls.'

Melville almost dared not to write his book, even as he advised a female friend not to read it. 'Dont you buy it – dont you read it, when it comes out,' he warned her. 'It is not a peice [*sic*] of fine feminine Spitalfields silk – but it is of the horrible texture of a fabric that should be woven of ships' cables and hausers. A Polar wind blows through it, & birds of prey hover over it.' With Mary Shelley's man-made monster at the back of his head, he conjured images of Ahab's ship ploughing through stormy seas as 'the ivory-tusked

Pequod sharply bowed to the blast, and gored the dark waves in her madness'. Only half jokingly, he spoke of his work as though it were some transgression of natural law which ought not to have appeared at all. 'But I don't know but a book in a man's brain is better off than a book bound in calf,' he told Evert Duyckinck, '– at any rate it is safer from criticism.' That binding might have been the tattooed skin of the pagan Queequeg; or the book his counter-bible, bound in the ghastly pale hide of the Whale itself. What began as an exercise in propaganda for the American whaling industry ended up as a warning to all mankind of its own evil. Melville had learned Hawthorne's lessons well.

It was, ostensibly, a cheery timetable, a rural regime. He rose at eight to give his cow and horse their breakfast before breaking his own, then settled to work till two thirty in the afternoon, when, by arrangement, Lizzie knocked and kept on knocking until her husband answered. After driving out in the countryside, he spent his evening 'in a sort of mesmeric state, not being able to read – only now & then skimming over some large-printed book'. Such self-imposed isolation seemed to invoke his increasingly strange and wilful voyage.

> I have a sort of sea-feeling here in the country, now that the ground is all cover with snow. I look out of my window in the morning when I rise as I would out of a port-hole of a ship in the Atlantic. My room seems a ship's cabin, and at nights when I wake up and hear the wind shrieking, I almost fancy there is too much sail on the house, and I had better go on the roof & rig in the chimney.

Working in the shadow of Mount Greylock, which he could see in the distance, the peak's blunt and sometimes snowy brow conjured up the White Whale, 'the mightiest animated mass that has survived the flood; most monstrous and most mountainous! That Himmelehan, salt-sea Mastodon, clothed with such portentousness of unconscious power.'

To friends, Melville spoke of the smooth running of his writing, but Lizzie wrote of a terrible time, a book accomplished 'under unfavourable circumstances – would sit at his desk all day not eating any thing till four or five o'clock – then ride to the village after dark'. Like Hawthorne – who walked around Concord with his head so bowed that he did not recognize buildings that he passed every day when he was shown photographs of them – Melville removed himself from human contact in order to write more forcefully about humanity. The result was a work written and performed in secrecy like a Masonic ritual, underlain with a conspiratorial text, what Melville told Hawthorne to be the secret motto of his book –

Ego no baptizo te in nomine patris, sed in nomine diaboli

– that is, 'I do not baptize you in the name of the Father, but in the name of the Devil'.

The Little Red Inn, Lenox, Western Massachusetts, 14th November, 1851, late afternoon, dreary snow and wind.

The chairs scrape over the boards as they draw nearer to the table. For two men to dine together was not usual in a country town. Melville had hired a private room for his publication party for *Moby-Dick.* There was only one guest.

Melville gave the finished copy to Hawthorne that afternoon. In those few seconds, as the book passed from hand to hand, between leaving go and taking hold, all the effort, all the energy of his life was distilled, the summary of his existence to date.

Hawthorne opened the book and saw the words inside:

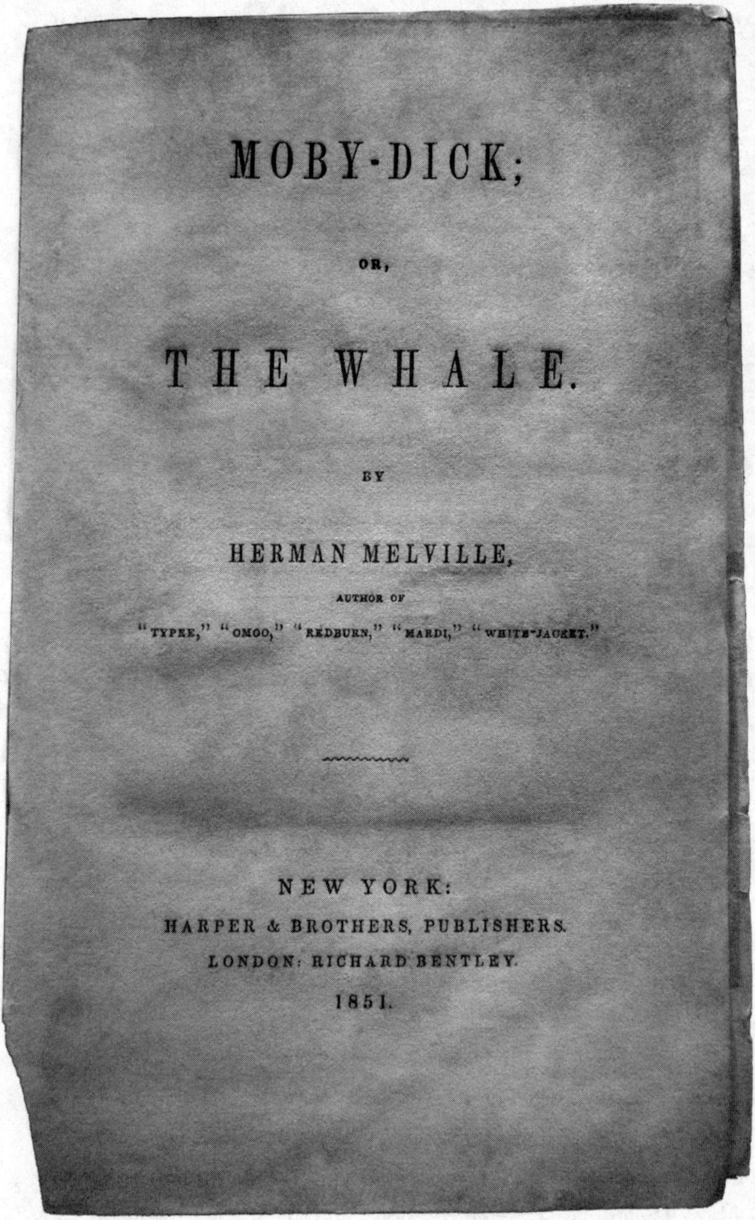

MOBY-DICK;

OR,

THE WHALE.

BY

HERMAN MELVILLE,

AUTHOR OF

"TYPEE," "OMOO," "REDBURN," "MARDI," "WHITE-JACKET."

NEW YORK:

HARPER & BROTHERS, PUBLISHERS.

LONDON: RICHARD BENTLEY.

1851.

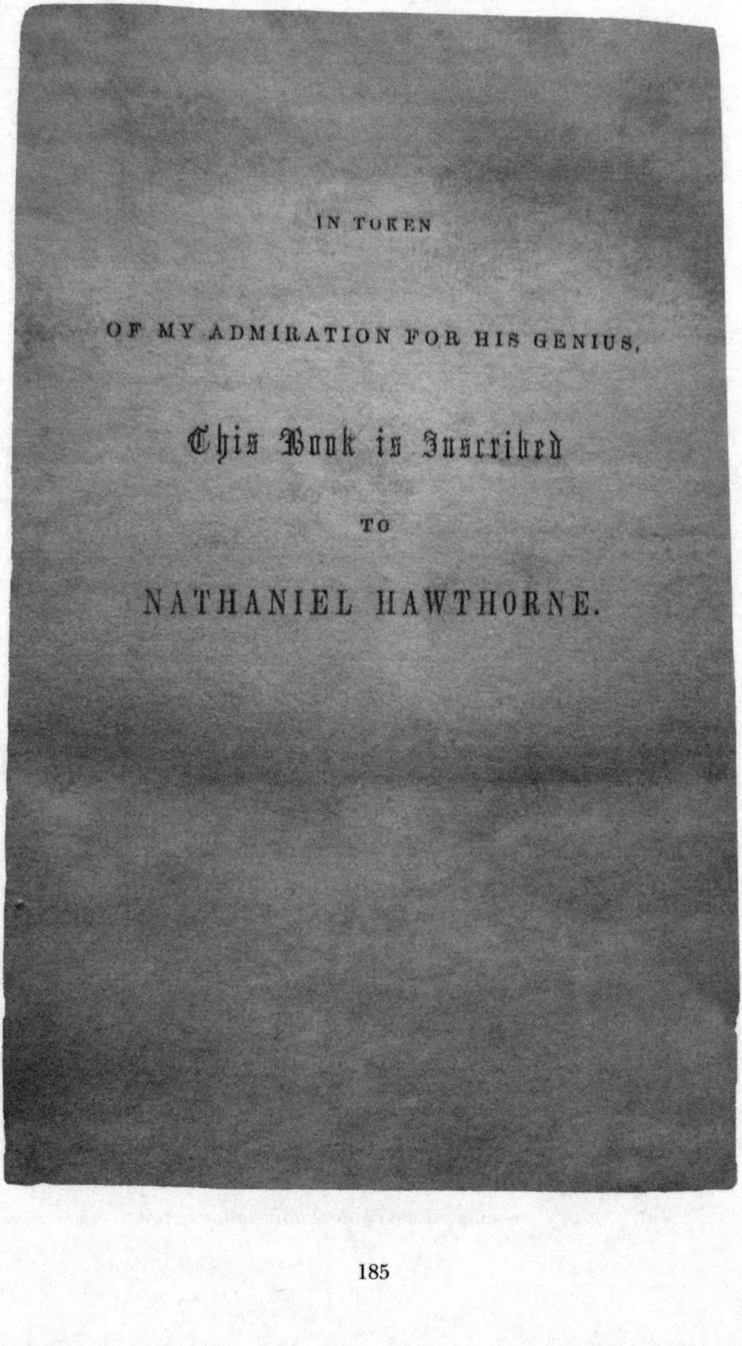

IN TOKEN

OF MY ADMIRATION FOR HIS GENIUS,

This Book is Inscribed

TO

NATHANIEL HAWTHORNE.

It was a public declaration, and an infinite demand.

Hawthorne's reaction to *Moby-Dick* is one of the great lost letters of literature, but we can see its shape by Melville's response.

> A sense of unspeakable security is in me this moment, on account of your having understood the book. I have written a wicked book, and feel as spotless as the lamb . . .

Hawthorne opened Melville's eyes to allegories and subtleties he had not seen in his own work. In response, and in an extraordinary mixture of arrogance and blasphemy and faith and love, the younger man almost accuses his friend and mentor:

> Whence come you, Hawthorne? By what right do you drink from my flagon of life? And when I put it to my lips – lo, they are yours and not mine. I feel that the Godhead is broken up like the bread at the Supper, and that we are the pieces. Hence this infinite fraternity of feeling . . . You understood the pervading thought that impelled the book . . . Was it not so? You were archangel enough to despise the imperfect body, and embrace the soul.

Even given the exaggerations of Victorian correspondence, these are dramatic words, and we can only imagine Hawthorne's reply. He may have been grateful he was about to leave Lenox. In Hawthorne, Melville sought refuge from the dark, like Ishmael and Queequeg settling down for their second night together at the Spouter Inn.

> Lord, when shall we be done changing? Ah! but it's a long stage, and no inn in sight, and night coming, and the body cold. But with you for a passenger, I am content and can be happy.

As his unholy book would be condemned by the good folk of the Berkshires, so he yearned for an eternity that his works, and those of his friend, might provide.

> I shall leave the world, I feel, with more satisfaction for having come to know you. Knowing you persuades me more than the Bible of our immortality . . . The divine magnet is on you, and my magnet responds. Which is the biggest? A foolish question – they are *One.*

It was a plea for fellow feeling that went beyond sex or intellect. It fed on the same unknowing power that drove his work; and as his relationship with Hawthorne could go no further – as he crossed the line of normal behaviour – so Melville never recovered from *Moby-Dick.*

On its publication, Melville's book confused and confounded the critics. Was it a gothic sensation, political parable, or a religious tract? Some thrilled to the chase, and the final battle between Ahab and the White Whale – 'he comes up to battle, like an army with banners . . . The fight is described in letters of blood' – but many were mystified, or even irate. Melville might have expected as much. He was more moved by the newspaper reports of a whale that had stove in a New Bedford ship. 'Crash! Comes Moby Dick himself, & reminds me of what I have been about for part of the last year or two,' he wrote to Evert Duyckinck. 'It is really & truly a surprising coincidence – to say the least . . . Ye Gods! what a commentator is this *Ann Alexander* whale. What he has to say is short & pithy & very much to the point. I wonder if my evil art has raised this monster.'

Despite its appearance on both sides of the Atlantic (like the White Whale, it could be in two places at the same time), the book prospered in neither. In order to register its copyright, it was first published in London under the title *The Whale*, in three volumes designed to catch the eye of the carriage trade, with bright blue boards and a handsomely embossed gilt whale swimming down each white spine. But just as that was a right whale – and therefore the wrong whale – so the expense of the English edition – which cost a guinea and a half, and which seemed to reflect the lavishness of that year's Great Exhibition – was undermined by Bentley's decision to excise the epilogue in which Ishmael survives to tell his tale (as well as sections he considered blasphemous or obscene), an omission that further confused the readers. The ending was restored for the American edition – a much more egalitarian, single volume affair, priced at a dollar fifty (although even this was available in a selection of differently coloured covers) – but Harper and Brothers never sold out of their three thousand copies, the remainder of which perished in a fire in the publisher's downtown Manhattan warehouse in 1853. It was a judgement, perhaps, to echo Hawthorne's bonfire of the vanities, and confirmation of its own author's assessment of his wicked book.

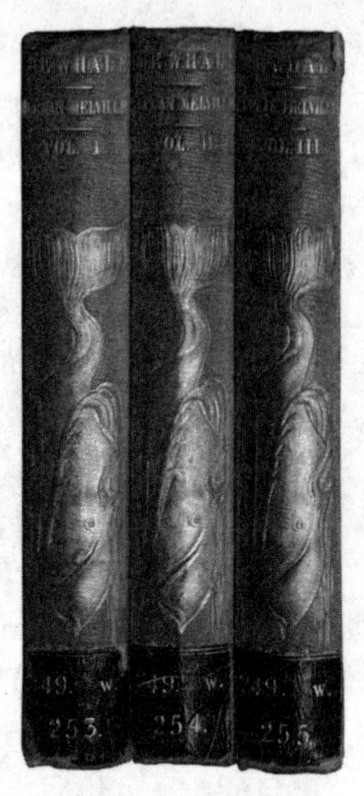

What made Melville also unmade him; it was the abiding paradox of his life. His adventures had provided him with material for his fiction, but they had ruined him for it, too, making him forever restless. By going to sea, Melville lived the life that would make his books possible; but his escapades also made him unfit for life as a writer. Haunted by the grand hooded phantom, the great whale, he felt dogged by 'the invisible police officer of the Fates, who has the constant surveillance of me . . . and influences me in some unaccountable way'.

Even as *Moby-Dick* was being published, Melville was at work on the decidedly land-locked *Pierre, Or The Ambiguities*, an autobiographical novel about a celebrated New York author who at one point is pursued down the street by a cameraman wanting to take his photograph, just as his alter ego had run from the Typees for fear their tattooists would take his face away. ('I respectfully decline being *oblivionated* by a Daguerretype,' Melville told another friend, 'what a devel of an unspellable word!') But his increasingly dark vision met with depressingly decreasing returns and a dwindling readership; and so in October 1856, despite suffering severe rheumatism, he embarked on what was to be his last great adventure.

'Mr Melville much needs this relaxation from his severe literary labours of several years past,' noted the *Berkshire County Eagle*, 'and we doubt not that he will return with renovated health and a new store of those observations of travel which he works so charmingly.' With him he carried his latest manuscript, *The Confidence-Man*, hoping to sell it in London. His ship arrived in Glasgow, where Melville marvelled at the shipyards and women with faces like cattle. At Edinburgh, he stopped to get his laundry done –

9 Shirts
1 Night shirt
7 Handkerchiefs
2 Pair stockings
Drawers & under shirt

– then proceeded, via Lancaster and York, to Liverpool, with its memories of his first sea voyage. Lodging at the White Bear on Dale Street, the next day he walked out in the rain 'to find Mr Hawthorne', but the address was out of date and his journey futile. The following morning he called at the consulate, and found Nathaniel.

Hawthorne had spent the last four years as American consul in Liverpool, living with his family in nearby Southport; he was now in his fifties, and balding. Melville too looked 'a little paler, and perhaps a little sadder'. Learning of his friend's ill health, Hawthorne diagnosed 'too constant literary occupation, pursued without much success', and a 'morbid state of mind . . . I do not wonder that he found it necessary to take an airing through the world, after so many years of toilsome pen-labour and domestic life, following upon so wild and adventurous a youth as his was.'

The two men took the afternoon train to Southport, a faded resort once patronized by Louis Napoleon, now a shadow of its former splendour. The next day they walked on the beach, blown along by the wind, and sat in a hollow in the dunes to smoke cigars. Melville began to talk of Providence and futurity, 'and of everything that lies beyond human ken'. He told Hawthorne that he had 'pretty much made up his mind to be annihilated'; like Ishmael leaving Manhattan, he seemed to advance a death-wish.

'It is strange how he persists – and had persisted ever since I knew him, and probably long before – in wandering to and fro over these deserts, as dismal and monotonous as the sand hills amid which we were sitting,' Hawthorne wrote in his journal. 'He can neither believe, nor be comfortable in his unbelief . . . If he were a religious man, he would be one of the most truly religious and reverential; he has a very high and noble nature, and better worth immortality than most of us.'

This was a high tribute from Hawthorne; a mirror, in its way, of the faith Melville had placed in him – as if only now he realized it and felt guilty for not having done more. But who could have saved Melville from himself? A few days later, he sailed from Liverpool for the Holy Land, leaving his trunk behind at Hawthorne's consulate, taking only a carpet-bag with him. The two men never met again.

Arrowhead is set close to the road, sheltered by trees. The rain washes the light out of the sky, the clouds rolling inky-black over the ochre house. Minutes later the sun is sharpening the clapboard, picking out acid orange day-lilies along the picket fence. Everything seems green and lush. Inside, the place feels uninhabited. Its wooden floors smell warm in the summer afternoon, but the rooms echo only to hushed voices. In the upstairs study, through the wavy, watery window, I can just make out the locked grey lump of Mount Greylock on the horizon, masked by trees.

. . . here and there from some lucky point of view you will catch passing glimpses of the profiles of whales defined along the undulating ridges. But you must be a thorough whaleman, to see these sights . . .

By the fireside is a toggle-head harpoon –

And if you be a philosopher, though seated in the whale-boat, you would not at heart feel one whit more of terror, than though seated before your evening fire with a poker, and not a harpoon, by your side.

– and nearby, a battered chest, left behind 'like a hurried traveller's trunk', with a handwritten luggage label, partly erased:

H. Melville – East 26th Street . . .

Our guide thinks Hawthorne was a handsome man, 'and that was the beginning of the trouble'. And I think of all those minor scenes, commonplace for all their protagonists' fame, two men smoking their cigars and drinking their brandy and staying up late, talking into the night.

> For now the words descended like the calm of mountains –
> – Nathaniel had been shy because his love was selfish –
>
> W.H. Auden, 'Herman Melville'

As dusk falls, the shutters come down. The doors are locked, and the house stands empty again. Mountains lie between – the mountain on which they met, and the mountain that marked their parting – rocks half covered in firs but bare to the summit, reaching out to the sky and back down to the sea.

It was the whiteness of the whale which appalled me

In 1863 Melville gave up trying to farm at Arrowhead and moved back to New York and a house in Gramercy Park. From there he would walk down to the Battery, where he earned four dollars a day as Deputy Inspector No.75 of the Custom Service, 'as though his occupation were another island.' In the evening he would work in his study, facing a wall like Bartleby. What did those years add up to but tragedy? In 1866, in the bedroom upstairs, his eighteen-year-old son Malcolm shot himself in the head with a pistol he kept under his pillow. Twenty years later, Stanwix, his other son, died of consumption, alone in a San Francisco hotel, aged thirty-four. As he looked through his window, across the street, Melville could see the terraced houses, mirrors of his own, their stone steps and iron railings a rhythm of urban banality, a view that never changed, unlike the sea.

His end would be as equivocal as his beginning. Melville was seventy-two years old when he died of a heart attack, just after midnight on a Monday morning in September 1891, before Manhattan had begun its working week. Thirty years had elapsed since his last novel, *The Confidence-Man*, and he had published only poetry since. After his interment in Woodlawns cemetery in the Bronx, Lizzie tidied up her husband's papers and put the manuscript of *Billy Budd, Sailor* away in a drawer. Glued to the inside of the desk on which he wrote it was a tiny clipping:

Keep true to the dreams of thy youth

Outside the city, in a bleak suburb – all the bleaker for a freezing February afternoon when the chill bleeds the colour out of the streets and sky – cars roar along the freeway in a twenty-four-hour race to get in and out of New York. They drive by without an upward glance to where their ancestors lie, having long given up the chase.

Shiny memorials line these tidy lanes; the names of city worthies are as deep-etched as the day they were set on these sepulchral avenues, suburbs of the dead, a sharp contrast to the simplicity of a Quaker graveyard. Last week's snow lies grey and gritty like an ice lolly spilt on the pavement. From my pocket I take a piece of slate, found on a Nantucket beach. I lean over to place it on the marble headstone, carved with ivy as if to mimic the living wreath growing around its feet.

HERMAN MELVILLE

Born August 1, 1819

Died September 28, 1891

Above the inscription is an extravagantly empty scroll, chosen by the author as his memorial; its blankness seems to mock all the books he did not write. Next to him lies Elizabeth, biding her silence, as ever; and on the other side, smaller memorials to his sons, both dead before their father. It is a sad array, a family reunited on a bare Bronx hill. Kicking at one of the little icebergs of frozen snow, I work up enough powder to shape a white whale on the lifeless grass, an acorn for its eye and a twig for its mouth. It looks childish, a cartoon animal playing over the writer's whitened bones. I wait to feel something, to commune with the writer's spirit. But there is nothing here, in this civic facility. The stone and the earth are all as dead as the asphalt over which the traffic hurtles en route for somewhere else.

BLACKFISH AT SO. WELLFLEET, MASS.
SOLD FOR $ 15000, DIVIDED AMONGST 300 INHABITANTS.)

VIII

Very Like a Whale

> Can he who has discovered only the values of whale bone and whale oil be said to have discovered the true use of the whale?
>
> Henry David Thoreau, *The Maine Woods*

From October 1849 to July 1855 – as Melville researched, wrote and published *Moby-Dick* – Henry David Thoreau undertook his walking tours of Cape Cod, having only recently emerged from his seclusion at Walden Pond, near Concord, where, in a two-year-long experiment, he sought to test the tenets of Transcendentalism.

The Transcendentalists, inspired by Ralph Waldo Emerson, sought a return to nature in order to feel God's true presence. Hawthorne saw them as 'queer, strangely-dressed, oddly-behaved mortals' – Victorian hippies, all but rehearsing Woodstock; they were satirized by Melville, too, for their romanticism: not least in the person of Ishmael himself. But for Thoreau, born in Concord in 1817, Walden was an escape from personal tragedy: the loss of his brother John, who had cut his finger when shaving and three days later died of lockjaw.

Walden was then still a wilderness, albeit one newly overshadowed by a railway embankment, built by the navvies from whom Thoreau bought his shack. Its sixty-one-acre pond is deeper in parts than Massachusetts Bay, with sandy shores shelving quickly to glacial black depths. Hawthorne found the water 'thrillingly cold . . . like the thrill of a happy death . . . None but angels should bathe there.' I saw no celestial beings when I swam there, but at the far end of the shore, in a glade beneath the pines and birches, there was a cairn of stones left by pilgrims to the site of Thoreau's hut.

Here, in a room under-tenanted by squirrels and racoons, the philosopher attempted a self-sufficiency of one. Here he recorded the minutiæ of the natural cycle, and his attempts to live with it. It was as if he had stalled his civilized life and re-geared it to natural forces. Like Hawthorne, who visited him there, seclusion charged his imagination. Thoreau revelled in the retreat of the day, and in the hours slowed by the calm surface of the water.

> As if you could kill time without injuring eternity.

He was almost childishly fascinated by the process of nature through which he hoped to examine the essence of existence. *Walden*, his account of those two years, is an alternative text for an industrial age, a kind of corollary to *Moby-Dick*. Axiomatic, philosophical, naïve and complex, it sometimes speaks with the voice of angels, sometimes with earthbound science. The writing of it is the true reason why Thoreau carried out his experiment, but that does not diminish its power. In his personal utopia, Thoreau sought to reinvent the way we could live. 'The mass of men lead lives of quiet desperation. What is called resignation is confirmed desperation.' He rejected the wisdom of the old – 'Age is no better, hardly so well, qualified for an instructor as youth, for it has not profited so much as it has lost' – and felt a sense of hubris in the manner in which he might mark his own immortality.

> What demon possessed me that I behaved so well?

Words came to Thoreau like a prophet of the new age, challenging the divisions wrought by his fellow men in their headlong pursuits. While at Walden, Thoreau protested against slavery and war by refusing to pay his taxes, a civil disobedience that earned him a night in gaol. Now aged thirty-two, and with only twelve years left before consumption took him, this man whom Hawthorne described as 'ugly as sin, long-nosed, queer-mouthed', yet whose character became him 'much better than beauty', had returned to Concord – barely two miles distant, yet a universe away.

With *Walden* published but, like *Moby-Dick*, hardly a success, Thoreau still felt the pull of nature, often travelling with his young cousin and intimate companion, Edward Hoar. Like Ishmael, Thoreau was drawn to the ocean. It was an irresistible lure for a loner – an '*Isolato* living on a separate continent of his own' – to seek out something greater and confront it; to seek refuge, too, from one's own self. The sea drew Thoreau out of the woods and onto the beach; the forest gave way to the ocean, the one opening out from the other. Yet neither was what it seemed, and like all desires, they were dangerous forces.

The Cape was barely more tamed than when the Pilgrims had made landfall there two hundred years before. Charles

Nordhoff, who visited it around that time, bemoaned the 'not over agreeable diversity of views' in the expanses of dunes, salt marshes, scrub oaks and stunted pines which earned it 'the euphonious name of "the Great Desert of Cape Cod."' It was certainly a sere landscape. 'Dreary-looking' wharves lined the bayside, while the stunted vegetation and absence of grass on the seaward side, 'and above all and mixed with all, the everlasting glare of the sand, all united to give the shores of the Cape a most desolate appearance'.

It was as dismal as the deserts Melville wandered in his mind, and Thoreau too found it a barren country, 'such a surface, perhaps, as the bottom of the sea made dry land day before yesterday'. Yet such bleakness also had its beauty: the high ridges of sand blown by Atlantic winds, over which were revealed the intense blue reaches; a mutability unwrought by man. This desolation – which Ishmael also saw in the limitlessness of the sea, 'exceedingly monotonous and forbidding' – appealed to the hermit of Walden; a place where 'everything told of the sea, even when we did not see its waste or hear its roar'.

Here the land paid homage to the ocean; became part of it, implicitly. 'For birds there were gulls, and for carts in the fields, boats turned bottom upwards against the houses, and sometimes the rib of a whale was woven into the fence by the roadside.' And here blackfish, or pilot whales, were prized for their oil, and had been since before the coming of the Pilgrims: the *Mayflower*'s second encounter with Native Americans had been at Wellfleet, where they watched Indians stripping the blubber from one of the stranded whales which earned it the name Grampus Bay.

Easily identified by their rounded melon heads and sleek black bodies, and so called because they followed a leader, pilot whales were hunted when other whales were not about. Frank Bullen recorded that 'a good rich specimen will make between one and two barrels . . . of medium quality', while hunks of their meat made a prized alternative to the ship's salt beef. These lithesome, lacquered cetaceans are, like their sperm whale cousins (with whom they often associate), highly social, and their propensity to gather in great numbers made them all the more attractive to catch. The people of the Faroe Islands still round up pilot whales using techniques learned by their Viking ancestors, driving entire schools into shallow water where, surrounded by small craft and men armed with all manner of weapons, the cornered whales leap and thrash, rising perpendicularly out of the water, as if straining every muscle to evade the deadly blades. Appallingly human in their physical presence, they might as well be men in wet suits, but they are soon reduced to butchered blubber.

Such scenes were played out on the Cape's shores, too. In an episode mirroring the striking first chapter of his book – which opens with the aftermath of a shipwreck and bodies being carried away in rough wooden boxes – Thoreau encounters slaughtered blackfish on the beach and is forced by the stench to take the long way round, only to find thirty more whales at Great Hollow, newly speared and turning the water red like the dead of a failed invasion.

Thoreau marvelled at the shape and texture of the animals, as smooth as India-rubber; with their blunt snouts and stiff flippers, they seemed almost embryonic. The largest was fifteen feet long; others were only five-foot juveniles with unerupted teeth, barely more than suckling babies. As the whales lay there, a fisherman obliged the visitor by slicing into the flesh to display the depth of the blubber, fully three inches thick. Thoreau ran his fingers over the wound, as if to believe. He felt its oily texture. It looked like pork to him; he was told that young boys would come along with slices of bread to make sandwiches of the stuff. The fisherman then dug deeper for the meat which, he told Thoreau, he preferred to any beefsteak.

As they stood there on the shore, Thoreau heard a cry: 'Another school.' In the distance, he could see the whales leaping through the waves like horses. The fishermen pushed off in their boats, boys running to join them. 'I might have gone too had I chosen,' said Thoreau; but he did not, nor was inclined to say why. Perhaps he felt the same equivocal fascination as I have when watching pink-coated huntsmen career through New Forest bracken. As Thoreau looked on, thirty boats rowed out either side of the whales, striking the sides of their craft and blowing horns to drive them onto the beach. He had to admit it was an exciting race, and as the frenetic scene played out before his eyes, he heard an old blind fisherman say, pathetically, 'Where are they? I can't see. Have they got them?'

For a moment it seemed the whales might win, as they headed north-west towards Provincetown and the refuge of the open ocean. Fearing their prey might be lost, the hunters were forced to strike then and there, using short-stemmed lances to take the whales as they leapt in and out of the waves. Thoreau could just make out the men as they jumped from their boats into the shallows, finishing off the animals as they lay on the beach, shuddering and spouting blood. 'It was just like pictures of whaling which I have seen, and a fisherman told me that it was nearly as dangerous.'

Those hunted whales haunted Thoreau. Back in Concord, he tried to find out more about them, but he discovered only an absence. Storer's *Report on the Fishes* did not include the pilot whale, 'since it is not a fish'; and Emmons's *Report of the Mammalia* omitted all whales, because the author had never seen any. I thought it remarkable that neither the popular nor scientific name . . . the Social Whale, *Globicephalus Melas* of De Kay; called also Black Whale-fish, Howling Whale, Bottlehead, etc., was to be found in . . . a *catalogue* of the productions of our land and water,' Thoreau mused.

It was a lack all the stranger for the part the whales played in the economy and history of the Cape: from the Indians' modest operations, to the modern 'early risers' who could still find one thousand dollars worth of whales stranded on the sand. Pilot whales and dolphins still strand here, in greater numbers than on almost any other shore. Lured by the presence of squid, the bay becomes a literal dead end for them, as they lie hoicked out of water, attacked by gulls which take advantage of the helpless animals to peck out their eyes as they slowly expire.

As he reached Provincetown, Thoreau marvelled at the part fishing village, part frontier town, with only one road and one pavement. 'The time must come when this coast will be a resort for those New-Englanders who really wish to visit the sea-side,' he predicted. 'At present it is wholly unknown to the fashionable world, and probably it will never be agreeable to them.' And as he approached his journey's end, Thoreau saw what looked like a bleached log on the beach. It proved to be part of the skeleton of a whale – a sign he conflated with a wreck that lay close by, its '*bones*' still visible: 'Perchance they lie alongside the *timbers* of a whale.' The Cape's winter storms still throw up eighteenth-century keels, their crossbars grey wooden ribs on the shore; but Thoreau could not know that these same sands also concealed a cetacean graveyard.

Dr Charles 'Stormy' Mayo is a man in his sixties, with a wiry frame, unflinching blue eyes, and a passion for growing dahlias. On his father's side, his family have lived on the Cape for nearly four hundred years; the Mayos first came to Chatham in 1650. His grandmother, on the other hand, came from the Azorean island of Faial. In his forebears' day, these waters were alive with animals, says Stormy, looking out of his office window and over to the bay. I can almost see the scene in his eyes, a paradise teeming with whales and fish.

Stormy's grandfather was one of the blackfish hunters – until the day he took a mother and heard her calf calling for her under his boat. He hadn't the heart for whaling after that. But he also told his grandson of a whaling station at the Eastern Harbor, on the outskirts of town, where the Cape is at its narrowest and most tentative. Had the sea broken through here, Provincetown would have become an island; but soon after Thoreau passed this way, a dyke was built across the slender stretch, and the harbour turned into a brackish lake.

And it was here, out walking, that Mayo and his son Josiah found a concavity in the dunes, a 'blow-out' that had temporarily ebbed to reveal a long-lost ossuary. Jaw bones and vertebræ lay jumbled together, sticking up out of the sand. Perhaps, like the elephants' graveyard, this was where whales went to die; whales once so numerous that the Pilgrims thought they might walk across the bay on their backs.

Lumbering and low, those whales' descendants still swim in Cape Cod Bay, labouring under their inauspicious name:

the *right* whale to catch; a ponderous, literal pun, borne with fortitude. With forty per cent of their body as fat, right whales are highly buoyant, spending most of their time at the surface; even more conveniently, they floated when killed. Along with their propensity to hug the shoreline – hence their other epithet, the urban whale – right whales suffered most of all from the centuries of unnatural predation. They were the first whale to be hunted, by Basques in the Bay of Biscay – a dubious honour commemorated by their proprietorial French name, *baleine de Biscaye* – but fewer than four hundred now remain in the North Atlantic.

With its baroque, glossy body encrusted with callosities, its paddle-shaped flippers and its bizarre, yawning mouth filled with baleen, *Eubalæna glacialis* is both grotesque and wondrous, the stuff of ancient engravings. It is the very definition of a whale, as supplied by Ishmael's sub-sub-librarian, who informs us that the word itself came of Scandinavian roots: *hvale*, meaning arched or vaulted in reference to its jaws, but also a reflection of the animal's rolling roundness, its architectural structure.

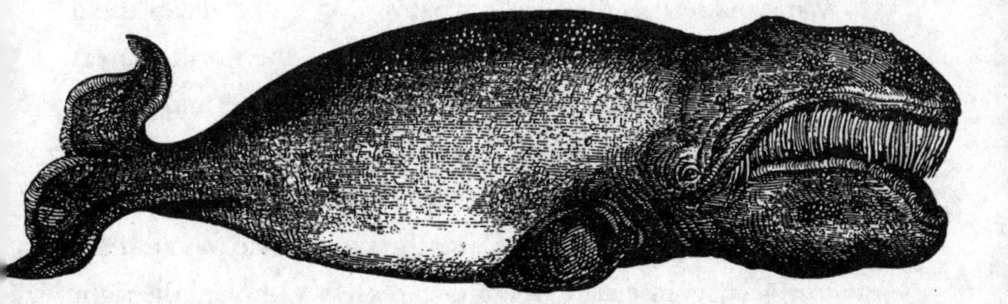

Like the sperm whale, the right whale was a victim of its strange physiology. Not only did it boast plentiful blubber, but its particularly long baleen, when heated, could be moulded into shape for umbrellas, corset stays and venetian blinds, or used as bristles for brushes. If whale oil was the petrol of its day, then whalebone was its plastic. Harvested in clumps higher than a man, their pliable blades were once arrayed in quayside plantations like giant sheaves of Jamaican sugar cane.

What made these the right whales to kill now makes them modern targets. Almost unbelievably, one of the world's rarest species chooses to frequent its most populous shores and busiest shipping routes. Here they fall victim to the tactics they deploy with their predators, remaining silent and still at the surface. An orca might be fooled into thinking the right whale was an inanimate object; an insensate freighter cares less. Although the right whale became the first cetacean to be protected from hunting in 1935, its numbers in the North Atlantic have remained static, despite legislation moving the shipping lane further north, and strict instructions that vessels should stay five hundred yards away from any whale. The result – so few breeding whales – is that the animal's gene pool is now so restricted that it is unlikely to survive the century.

The irony is that the right whale is such a fertile, if not fecund creature. Weighing nearly a ton, the male's testes are the largest of any animal. These, along with its eight-foot penis, allow it to take part in sperm competition in which males assert their supremacy by multiple matings rather than fighting for favours (although they may use their callosities as a kind of weapon). Females will even permit more than one partner to enter them at the same time, after sessions of delicate foreplay in which the courting animals use their flippers to stroke each other with inordinate gentleness; like all whales, their skin is incredibly sensitive, and the pressure of a human finger can send their entire body quivering. Despite this vigorous approach to sex, there are only eight matrilineal lines left in the northern species – the visible legacy of centuries of whaling.

Stormy saw his first right whale when he was a sixteen-year-old boy out fishing on Stellwagen Bank with his father; they were almost legendary animals by then, already close to extinction. 'People knew there were some left,' he recalls, 'but nobody knew where most of them were.' His youthful interest evolved into an adult passion, and having helped found the Center for Coastal Studies in Provincetown in 1976, Mayo began collecting data on right whales. He also became the first person to be licensed by the government to rescue entangled whales; more than sixty per cent of right whales have been caught in fishing line. Ship strikes, too, are killing more than hitherto suspected, many of them females of calf-bearing age. At this stage of the species' history, to save just one fertile female could make a difference between extinction or survival. It is hard not to see Stormy and his colleagues as new heroes of anti-whaling, flown interstate at short notice on operations costing thousands of dollars.

Indeed, the same techniques that were once used to hunt whales are now used to save them. Alerted to an incident, the *Ibis* reaches the scene as soon as possible; an entangled whale may eventually die from starvation or infection, but more immediately, it can drown. The rescuers use a rigid inflatable boat to approach the animal and attach sea anchors to slow its progress, just as whalers used wooden kegs, and Native Americans attached inflated sealskins to harpooned whales. Wearing an ice hockey mask and a helmet (on which is fixed a video camera to record the event), Stormy attempts to cut the victim free. His equipment may be twenty-first century, but in silhouette he resembles a Victorian harpooneer, only instead of a barbed spear, he wields a long-handled hook to slice through the cat's cradle of line.

As Stormy bears witness, an angry whale is a dangerous whale. These are not the equivalent of cetacean cows grazing over verdant oceanic pastures (although they are closely related to ruminants, and have multiple stomachs to digest their food). Rather, they are surprisingly flexible creatures – much more so than other baleen whales, despite being twice as heavy as humpbacks, and positively barrel-shaped compared to the streamlined but largely unbending finbacks – able almost to touch their flukes with their snouts, an act of acrobatics necessary for turning in tight circles when pursuing their minute, ever-shifting prey.

On his computer, Stormy runs video clips of entanglement scenes. The sheer muscular power of the animal is vividly apparent. Twisting and turning like a gigantic salmon, the whale's tail thrashes in a manner that brings those nineteenth-century scenes alive. With one flick of the tail, this wilful creature could truly send a boat flying into the air, 'his very panics . . . more to be dreaded than his most fearfulness and malicious assaults!' as Ishmael observes.

Stormy's relationship with the right whales is intimate by virtue of such close encounters. He speaks of the prehistoric vision they present, the sun glinting through their baleen. And while he finds the use of the word 'intelligent' less than useful in conjunction with animals, he does not hesitate in calling them 'wicked', creatures that know their power. The whalers knew this well. Unable, like the sperm whale, to see straight ahead or behind, the right whale could 'sweep with his tail or flukes from one eye to the other, thus rendering any approach to his body, from abreast, impossible or highly dangerous'.

After a rescue, Stormy often cannot remember what he has done. He reasons this is because his short-term memory deals only with essential details; only when he watches the playback does he relive the moment. Once, a fishing hook on the line snagged on Stormy's lifejacket as the animal dived, threatening to take him down too, like Ahab tethered to Moby Dick. Mayo had split-seconds to cut himself free; in the water there would be no escape as the momentum dragged back his arm, making it impossible to use his knife.

> . . . And were the whale then to run the line out to the end in a single, smoking minute as he sometimes does, he would not stop there, for the doomed boat would infallibly be dragged down after him into the profundity of the sea . . .
>
> The Line, *Moby-Dick*

On his computer, Stormy's colleague, Scott Landry, shows me other images of entanglement: animals with nylon line cutting so deep that the flesh has begun to grow over it, even as it weeps and bleeds; whale lice or cyamids colonize these weakened areas, signs of an ailing animal. It is upsetting to see sleek bodies turning ghostly grey, sapped by the cords that bind them. By-products of global endeavour rather than subjects of it as they once were, they must have sinned mightily to be so ill rewarded by fate. A last picture shows a dead whale on a beach, livid and pink, visibly diminishing as a recognizable creature, although its eye still stares and weeps.

> We need another and a wiser and perhaps a more mystical concept of animals . . . We patronise them for their incompleteness, for their tragic fate of having taken form so far below ourselves. And therein we err, and greatly err. For the animals shall not be measured by man. In a world older and more complete than ours they move finished and complete, gifted with extensions of the senses we have lost or never attained, living by voices we shall never hear. They are not brethren, they are not underlings; they are other nations, caught with ourselves in the net of life and time, fellow prisoners of the splendour and travail of the Earth.
>
> Henry Beston, *The Outermost House*, 1926

In the late winter and early spring the Center's research vessel *Shearwater* sails out to measure the levels of zooplankton in the bay. The theory is that these levels are accurate indicators of whether this habitat can support the whales. If the count rises above 3,750 organisms/m^3, then the density of oil-rich copepods and other colourless animals – each looking, under the microscope, like little watery extraterrestrials as they row themselves in eccentric circles – will sustain the population. If not, any whale calling on this historic feeding ground will find it wanting and move away. From such minute and methodical study, leviathans follow.

Zipped and velcro'd into a padded, astronaut-like survival suit to forestall my death from hypothermia should I tumble over the side of the boat, I sign away any claims to public liability and, duly approved by the federal government, I climb the metal ladder to the *Shearwater*'s upper deck, facing the bright sun and chill wind. Despite instructions on how to focus just below the horizon and see with my peripheral vision, the unchanging surface and the sea's motion lulls me into a kind of sleep. 'There you stand, lost in the infinite series of the sea, with nothing ruffled but the waves,' as Ishmael says, watching for whales from the mast-head, 'everything resolves you into languor'.

Nothing breaks the monotony, not even a bird. It is as if the whole world had been chilled. After six hours' searching, my eyes begin to ache. Everything is flat, almost soporific in the icy winter air. The long windsock-like sample nets trailing emptily from the stern prove a negative: there is not enough food here for whales.

It is not a good sign. Cold water holds more oxygen, and so supports more food than southern seas, but rising temperatures have driven plankton north by a latitude of ten degrees, while warming oceans absorb more carbon dioxide, acidifying the whales' environment. The spotter plane circling overhead sees not so much as a blow, and the sun and wind burn my face for looking. Perhaps we were just not worthy that day.

* * *

Three months later, I sailed on the *Shearwater* again. It was early May, and the right whales had not appeared in any numbers; the plankton counts remained frustratingly low. But then Stormy reported a change in the circumstances.

From Provincetown's harbour, the boat made for the western side of the bay, eight or ten miles towards Plymouth. We sat on the upper deck, watching porpoises slip through the water, fleet and shy; it was easy to see why sailors called them sea pigs as they snorted and shuffled through the waves. Then we saw something else: a low dark shape gliding along the surface. It seemed almost inconsequential; but as we drew closer, I realized it was a right whale. Slowly but surely, the animal was moving like a lawnmower, purposefully harvesting the now plankton-rich waters, called here by some collective memory, or perhaps by smelling or even hearing its food. As the *Shearwater* closed the distance between, I put down my field glasses and looked on in amazement.

One, two, three, four, five whales now appeared around us, baleen plates glinting in the sun like enormous musical instruments. Suddenly their incongruous beauty was revealed, the strange bonnet at the top of their heads, covered in pale growths like lichen on a tree. As they floated, buoyed up by their bulk, they looked more like plants than animals, or maybe shiny rocks, kept glossy by the water running over them. Only behind and below was their power evident, their broad flukes barely breaking the surface, effortlessly manœuvring their bodies.

They were giant, living jigsaw puzzles: no matter how hard I looked, I could not grasp the entirety of the creatures, the sense of their structure, the components from which they were made. It was as if they were shifting in and out of focus. As we came up

. . . even so these monsters swam, making a strange, grassy, cutting sound; and leaving behind them endless swaths of blue upon the yellow sea . . .

behind one animal, I saw how broad was its back; how it shelved out from its spine like a great table, and I could imagine why Brendan the navigator and his monks landed on a whale and, presuming it to be an island, lit a fire and said Mass in thanks for their salvation.

Abruptly, one animal approached the boat, so close that Stormy – who was standing on the bowsprit, held out over the waves – could have reached out and patted its rough head like a dog. Instead, he focused his camera to record the pattern of callosities, which appear at the same points where hair grows on the human face – brow, chin, upper lips – and which give each animal its identity, a rough physiognomy queasily underlined by the fact that they are infested with pale whale lice, the minute scorpions that crawl around their host's head, eating its dead skin. As Eric Joranson, one of the mates on the whale watch boats told me, the lice will also colonize a human given half the chance, and are difficult to dislodge once they do. When a whale lies dying on the beach, they leave it like rats off a sinking ship. None the less, these parasites may also assist the whale: since they eat the same copepods, it is possible that they lead it to its food, acting as minute sensors.

As the whale passed us, it was as if it were paying court to its champion, nodding its head serenely towards Stormy as it passed. It then swung around the boat, and next to me. Looking down into the water, I could see its great white jaw swinging open like some massive hinged door, wide enough to garage a car – the largest mouth of any living creature. Now I could see the entirety of the animal, hanging below, an iceberg suspended in its element. It was also deceptively fast, creating a wave in front of its snout with the weight of its fifty tons. Silently gaping as it passed by, both aware and unaware of us, it was like watching a dinosaur, an animal whose physical presence was belied by its air of fatality. It also smelled, a deep insupportable smell, somewhere between a cow's fart and a fishy wharf, a pungent reminder of its function as a processing plant for plankton.

Then it was gone to join the others, apparitions that, for all their size, were quite dreamlike. It was hard to look on these huge creatures and think of a time when they might not be there. Barely a mile away, shipping was moving in and out of Cape Cod Canal and under the distinctive hump of the Sagamore Bridge. It was a lesson in the nature of survival. Paying scant attention to anything other than their food and themselves, they would not know, could not see, the tanker or the container ship steaming towards them. Later that day, the *Shearwater* alerted shipping to their presence in the bay. What was a day trip for me may have saved a whale's life.

As we turned to leave, a black shape broke the horizon. A whale was breaching, lazily launching itself into the air before landing with a distant crash. Then it began to slap down on the surface with its tail, the sound ricocheting off our boat as a cannonade. As it held its flukes emblematically against the sky, infused with its own life and power, we turned our backs on the whales and left them to their lunch.

> *Ham.* Do you see yonder cloud, that's almost in shape of a camel?
>
> *Pol.* By the mass, and 'tis like a camel, indeed.
>
> *Ham.* Methinks, it is like a weasel.
>
> *Pol.* It is backed like a weasel.
>
> *Ham.* Or, like a whale?
>
> *Pol.* Very like a whale.
>
> *Hamlet,* Act Three, Scene II

Hamlet was right, for all his teasing. Whales are like clouds. They change shape, forming and re-forming as they pass through the great expanse of the sea and over its drowned mountains and valleys, just as the clouds drifted over the snowy peak that Melville watched from his window in a Massachusetts meadow. In whale bone carvings, the Inuit represent the whale's breath as a feather. Cartoon whales spout their own personal weather, their own head of steam. To its prey, the white belly of a humpback, too, appears as a cloud, albeit one that might consume it.

And as clouds create atlases in the air, so whales are countries in their own right, planetary communities of barnacles and sea-lice wandering on their own continental drift. International ambassadors of nature's undiscerning power, they are stateless nations, invested with something beyond their mere presence. 'By art is created that great Leviathan,' wrote Hobbes, 'called a Commonwealth or State.' As plundered colonies, they remain under attack, invincible yet vulnerable, defenceless for all their size. It is the whale's fate to share man's air, and so risk its life in the process of sustaining it, caught in a bind as much as any philosopher perplexed by the human condition.

The whale lives between worlds; that is its miracle, and its folly. What did it do to deserve such a fate? Spurned by Noah (it could hardly fit in the ark), it pays the price for its self-exile, having forsaken the land for the sea.

> I am, by a flood, borne back to that wondrous period, ere time itself can be said to have begun; for time began with man.
>
> The Fossil Whale, *Moby-Dick*

The earliest whale-like creatures can be traced back fifty million years to the Eocene and the Tethys Sea, an ancient ocean whose vestiges now form the Mediterranean and Caspian seas. Their ancestors included *Pakicetus*, a four-legged and fox-like creature which in turn gave way to *Ambulocetus natans*, a kind of giant otter, and other so-called 'walking whales' such as *Kutchicetus* and *Rodhocetus*. Recent discoveries point to another missing link between whales and land-dwellers: *Indohyus*, a deer-like ungulate which possessed a similar bony structure in its auditory system to that of cetaceans; being a herbivore, it became semi-aquatic to escape its predators. Drawn to the water's edge, the mesonycids' descendants would become horses, bison, camels, sheep – and cetaceans.

The first whales, the archæocetes, were quite as global in range as their descendants – although the serpentine remains of *Basilosaurus cetoides* convinced Victorian palæontologists that it was a marine reptile when its fossilized skeleton was found in the Deep South in 1832; Ishmael claims that 'awe-stricken credulous slaves in the vicinity took it for the bones of one of the fallen angels'. Only Sir Richard Owen, inventor of the dinosaur, recognized this 'annihilated anti-chronical creature' as a 'pre-adamite whale' which he renamed *Zeuglodon*, 'one of the most extraordinary of the Mammalia which the revolutions of the globe have blotted out of the number of existing beings'.

Around thirty-five million years ago, the whales divided into mysticetes and odontocetes, leaving the archæocetes to become extinct; although some scientists believe that sperm whales are genetically closer to baleen whales than they are to other toothed whales. Similarly, recent fossil discoveries of the baleen whales' antecedents have revealed animals with huge, baleful eyes and jagged teeth, quite unlike their benevolent, modern counterparts.

Given the lacunæ in the fossil record and our ignorance of great swathes of time, the evolution of the whale remains obscure. Traces of their land-borne origins can be seen in the residual hind limbs of embryo whales, as if their prehistory could be read there – but then, we are all whales in the womb, swimming in amniotic seas. Occasionally a sperm whale is born with an extra pair of atavistic fins, while one humpback was recorded with freak limbs a yard long, a strange being, neither one thing nor the other, like a Barnum mermaid made from a fish and a monkey.

The whales made good use of their freedom from the land. It is the buoyancy of the sea that has allowed them to develop into such mighty animals: if they still had legs, they would not be able to stand on them, so great is their weight. Such an evolutionary genesis both refutes and reflects the hand of the Almighty: as one Victorian handbill, advertising the exhibition of a whale's bones, claimed:

> Who can contemplate this mighty skeleton . . . without adoring the Mind that formed it? Where can we better cultivate a sentiment of devotion than in the presence of work so expressive of the various attributes of the varied God?

Yet for an era whose beliefs were under threat, the whale had a kind of equivalence with the origins of the earth and the newly discovered animals of prehistory; if these cetacean giants survived the flood, then so might other monsters. 'Leviathan is not the biggest fish,' as Melville told Hawthorne, '– I have heard of Krakens.'

In the first half of the nineteenth century sea serpents were sighted, with remarkable frequency, off the coast of

Massachusetts. Witnesses claimed to see huge animals with snake-like bodies and heads held high out of the water. Unlike many such fantastical monsters, however, these beasts were seen by hundreds of people for hours at a time, and no less a body than Boston's Linnæan Society published its findings on the subject in a pamphlet, a copy of which is lodged in the British Library, stamped with the name of its owner, the naturalist Joseph Banks.

'In the month of August 1817, it was currently reported on various authorities, that an animal of very singular appearance had been recently and repeatedly seen in the harbour of Gloucester, Cape Ann, about thirty miles distant from Boston,' noted the society, whose members were all Harvard graduates and one of whom, Jacob Bigelow, was a renowned scientist and inventor of the term 'technology'. 'It was said to resemble a serpent in its general form and motions, to be of immense size, and to move with wonderful rapidity; to appear on the surface of the water only in calm and bright weather; and to seem jointed or like a number of buoys or casks following each other in a line.' In response, the society appointed a committee 'to collect evidence with regard to the existence and appearance of any such animal'. It was a court of law called to consider the existence of sea monsters; and although its findings might have certified a hippogriff, it is hard to conclude that the witnesses did not see what they say they saw.

Amos Story of Gloucester, mariner, said the animal had a head shaped like that of a sea turtle, 'his colour appeared to be a dark brown, and when the sun shone upon him, the reflection was very bright. I thought his body was about the size of a man's body.'

A Monstrous Sea Serpent,

The largest ever seen in America,

Has just made its appearance in Gloucester Harbour, Cape Ann, and has been seen by hundreds of Respectable Citizens.

Solomon Allen of Gloucester, shipmaster, saw it three days running, 'nearly all day from the shore . . . I was on the beach, nearly on a level with him . . . He turned short and quick, and the first part of the curve that he made in turning resembled the link of a chain.'

Epes Ellery of Gloucester, shipmaster, witnessed 'the upper part of his head, and I should say about forty feet of the animal . . . I was looking at him with a spy-glass, when I saw him open his mouth, and his mouth appeared like that of the serpent; the top of his head appeared flat . . . He appeared to be amusing himself though there were several boats not far from him.'

In their deliberations, the committee consulted historical accounts such as that of Bishop Pontoppidan's *Natural History of Norway* of 1755. The cleric recorded that experienced seamen found it strange to be asked if such creatures existed; he might as well have asked if there were such fish as cod or eel. Bearing such evidence in mind, the Linnæans declared 'the foregoing testimony sufficient to place the existence of the animal beyond a doubt'.

It was a remarkable conclusion, and as if to mark it, a second serpent was seen in Long Island Sound that October, 'perhaps not more than a half mile from the shore, a long, rough, dark-looking body, progressing rapidly up sound (towards New York)'. One witness watched through his telescope as its back, forty or fifty feet of which was visible, rose above the surface, 'irregular, uneven, and deeply indented'. It was a somehow horrifying scene, of a monster approaching Manhattan, and was revisited later when another was seen eighty miles up the Hudson River. A further sighting off Nahant, Boston, was witnessed by at least two hundred people.

Over the following years, these creatures reappeared in the same waters, as though summoned by the same upwelling of food that brought the whales to the Gulf of Maine. In May 1833, for instance, five officers of the British garrison out fishing for the day in Mahone Bay off Halifax were surprised by a school of pilot whales 'in an unusual state of excitement and which in their gambols approached so close to our little craft that some of the party amused themselves by firing at them with rifles'.

Only then did the officers realize that the whales were fleeing 'some denizen of the deep' two hundred yards behind. Its movements were 'precisely like those of a common snake, in the act of swimming, the head so elevated and thrown forward by the curve of the neck, as to enable us to see the water under and beyond it'. The creature was estimated at one hundred feet in length.

That August, the British consul watched a similar animal from a hotel terrace in Boston: 'above a hundred persons saw it at the same time'. One was even seen at Herring Cove in Provincetown, apparently enticed by the presence of fish and warmer waters. No less a person than Senator Daniel Webster saw a monster off Plymouth, a sighting recorded in Thoreau's *Cape Cod*, along with the politician's urgent request to fellow anglers never to mention a word of the encounter, lest he spend his life being asked about it. Little wonder that the serpent was a topic of conversation at the luncheon party after Melville and Hawthorne met on Monument Mountain; or that down in Carolina, another monster caused a sensation when it swam up the Broad River into one of its tributaries which was barely a hundred yards wide, chased all the while by a party of men shooting at it with rifles.

Throughout the century, from all corners of the globe, there were sightings of sea serpents. Surely not even the sceptical could dismiss them as a conspiracy of fools? Spectators swore to the same details: a huge, long-necked animal, able to swim faster than the fastest whale. Precise locations were given, in latitude and longitude, marking these appearances at exact moments in time, to be entered in ship's logs, and relayed in newspaper paragraphs. Such was the evidence surrounding the sea serpent that when Henry Dewhurst published his *Natural History of the Cetacea* in 1834, he included it as fact, 'one of those unknown animals which occasionally puzzle the zoologist when they make their appearance'.

Scanning the yellowing columns of newsprint, it is remarkable to see how often such mythical animals rear their head, and what a debate raged around the possibility of their existence. The most famous encounter came on 6 August 1848, when the crew of HMS *Dædalus*, en route from the Cape of Good Hope to St Helena, watched 'an enormous serpent, with head and shoulders kept about four feet constantly above the surface of the sea'. Sixty feet of the animal was visible *à fleur d'eau*, as Captain McQuhæ, stirred to poetic French by the apparition, described it. The creature (which to me looks rather like a giant slow worm) passed so close that had it been 'a man of my acquaintance', the captain added, 'I should have easily recognized his features with the naked eye'. Readers of the *Illustrated London News* thrilled to a double-page spread on the subject, along with the testimony of an officer of the Royal Navy.

But of all these accounts, it is those that describe interaction with whales that most terrify and intrigue; not least because, in such company, they seem to disprove the assertions of experts who claimed that what experienced sailors saw were in fact whales, sharks, porpoises, or even elephant seals. In June 1818 eighteen passengers and the captain of the packet *Delia*, sailing off Cape Ann, watched a sea serpent battling a humpback whale, the creature rearing its head and tail twenty-five feet out of the water. In July 1887 a monster was seen fighting what was presumed to be a cetacean off the Maine coast; the following morning, a dying whale was found stranded nearby, 'its flesh torn and gashed'. The most extraordinary report, however, came from the South Atlantic in 1875.

On 8 January, off Cape São Roque, on the north-eastern corner of Brazil – a landmark for emigrating whales – the barque *Pauline* was sailing in moderate winds and fine weather when her crew saw some black spots on the water, with a whitish pillar high above them. As the ship drew near, it became apparent that the pillar was more than thirty feet tall, and was rising and falling with a splash. George Drevar, ship's master, picked up his eye-glasses and could not believe what he saw: a sea serpent with its coils wrapped twice around a sperm whale.

At this point in the narrative I find it almost impossible to proceed, for fear of waking the sleeping monster, wondering if I'd ever dare go out of my depth again.

Using its head and tail as levers, the serpent was twisting itself around the whale 'with great velocity'. Every few minutes the pair sank beneath the waves, only to reappear, still engaged in mortal combat. The struggles of the whale – along with two others nearby which were 'frantic with excitement' – turned the sea around them into a boiling cauldron, rent with loud and confused noise. From its coils, Drevar estimated the serpent to be in excess of 160 feet long. He noted that its mouth was ever open, an observation that somehow makes the scene more awful. As the crew of the *Pauline* watched, the battle of the leviathans continued for fifteen minutes, and ended only when the whale's flukes, waving backwards and forwards and lashing the water in its death throes, vanished below, where, Drevar had no doubt, 'it was gorged at the serpent's leisure; and that monster of monsters may have been many months in a state of coma, digesting the huge mouthful.'

With that final act, the two sperm whales that looked on, 'their bodies more than usually elevated out of the water', moved slowly towards the ship, as if seeking shelter from the monster. They were 'not spouting or making the least noise, but seeming quite paralysed with fear'. The engraving made from Drevar's sketch merely underlines the pathos: the cruel sea serpent playing with the placid whale like a cat with a garden bird, twisting and turning it in its grasp as the cetacean fights for its life.

It may be that Drevar actually witnessed the titanic struggle between a giant squid and a whale. I must confess I have seen whales that look like sea monsters, rolling in the waves. My childish desire to believe in a lost world (Arthur Conan Doyle, on honeymoon in Greece, claimed to have seen a young ichthyosaur in the sea) seeks to create something palpable out of the apparently incredible; to conjure an abyssal nightmare out of the pages of scientific certainty. Yet fishermen, clergymen and men of experience and social standing risked ridicule by swearing to what they saw. Could they really have been deceived by schooling porpoises or basking sharks?

Doubt would surround the sea serpent until the day it was caught and presented for public display. In 1852, a year after the publication of *Moby-Dick*, a New Bedford whaler promised to do just that. Sailing in the South Pacific, the *Monongahela* claimed not only to have seen a sea serpent, but to have pursued and harpooned it like a whale. The 103-foot-long animal was brought on deck, dried and preserved, along with its long, flat, ridged head and ninety-four teeth, 'very sharp, all pointing backward and as large as one's thumb'. This remarkable finding, which seemed set to prove the existence of the beast, was reported in the British journal *Zoology*, after the whale-ship had gammed with a brig which brought back the captain's letters describing the monster. But the *Monongahela* never reached her home port. A year later she was lost at sea with all hands, and her incredible cargo. What a specimen it would have made for New Bedford's museum and Ishmael's eyes: the great sea serpent on display.

In one of his most mystical asides, 'A Bower in the Arsacides', Ishmael tells of an exotic island, supposedly in the Mediterranean, where a whale's skeleton had become a place of worship. Its ribs were hung with trophies, its vertebræ carved with a calendar, and in its skull burned an eternal flame, 'so that the mystic head again sent forth in vapory spout; while . . . the wood was green as mosses of the Icy Glen . . .' In this living temple of growth and decay, the vine-clad bones turned into a verdant bower – 'Life folded Death; Death trellised Life' – and our narrator takes the opportunity to have the dimensions of this Arsacidean whale tattooed on his own body, 'as in my wild wanderings at that period there was no other secure way of preserving such valuable statistics'.

To Ishmael, the whale is as mysterious as any sea serpent: a formidable creature to be feared and even worshipped. And as this gothic episode – with its evocations of the gloomy glade where Melville and Hawthorne met – draws Ishmael's attention across the Atlantic, so his report summons me home, to discover what became of the whale in my native land. For it was in England that the true nature of the leviathan would be made known; from English whaling ports that distinguished men would set out to identify, categorize, and perhaps even pin down for posterity the still somewhat conditional reality of the whale.

IX

The Correct Use of Whales

> There is a Leviathanic Museum, they tell me, in Hull, England, one of the whaling ports of that country, where they have some fine specimens of fin-backs and other whales . . . Moreover, at a place in Yorkshire, England, Burton Constable by name, a certain Sir Clifford Constable has in his possession the skeleton of a Sperm Whale . . . articulated throughout; so that, like a great chest of drawers you can open and shut him, in all his bony cavities – spread out his ribs like a gigantic fan – and swing all day upon his lower jaw.
>
> A Bower in the Arsacides, *Moby-Dick*

Even its name sounds inexpressive, yet more so in the dialect of the flat east Yorkshire coast, barely a word at all: *'ull.* But seen from the suspended bridge that sails over the Humber before it reaches the grey waters of the North Sea, the city aspires to its proper name: Kingston-Upon-Hull, a pride evinced in its cream-painted and crowned telephone boxes – a defiantly independent network for an imperial place.

As you descend to the banks of the estuary, the industrial sprawl becomes evident; factories compete with retail sheds to brutalize the landscape. They cannot quite destroy the impression, so carefully constructed by the civic forefathers, of an age of trade and certitude; of an affluence set in sandstone and grand municipal works. In the city centre, at the end of a narrow street, is the gabled home of Hull's favoured son: William Wilberforce, liberator of slaves and founder of the Society for the Suppression of Vice. Nearby stands a giant column, surmounted by a statue of the man, broadcasting his achievement in capital letters –

NEGRO SLAVERY
ABOLISHED
1 AUGUST
MDCCCXXXIV

– although it faces another building, one that belies the city's claim to manumission.

Passing through the double doors with their polished brass fingerplates, I follow a sombre corridor over which hangs a skeleton, showing the way, just as I can hear a strange sound rising to a stifled focus, sometimes like a choirboy, sometimes like a trapped dog, luring me on, just as all the sounds I ever heard are compressed into one continuous noise in my head. The lighting is barely brighter in the room beyond. There is a bench, although not for public use. It is built from the bones of a whale: blade bones for the seat, ribs for the back and arms. Next to it is a hat stand created from a narwhal's tusk nailed to a wooden base.

On the far wall of this macabre salon hang a pair of portraits both labelled, confusingly, William Scoresby. In the first, a rotund man points over a homely cottage to a ship in the distance; with his white waistcoat, belly and ruddy face, he might be a bluff farmer, rather than a reaper of the sea. The second picture shows his son in starched collar and stock; his are the refined features of a man of the Enlightenment. Between them, these two Scoresbys – one a lifelong merchantman, the other destined to be a fellow of the Royal Society – preside over a collection whose enthusiasm for its subject has faded over the years, like a stamp album put away in the attic, partly from embarrassment at such youthful and compulsive fervour.

The museum's displays are contrived to resemble a ship's superstructure. Everything seems subfusc. Set into the bulwarks are framed photographs, backlit to give them life, although one might almost wish they weren't. Sepia ghosts projected out of the ether, Yorkshire's hardy sons labour in the Arctic, in scenes as industrial as any in Bradford's mills. Tall ships stand glamorously rigged in the ancient sunlight, while their workers' faces stare out, stilled in the moment.

Above them, curious souvenirs are caught in the cordage. Hauled up the mainmast is the sleek carcase of a narwhal, its leopard spots losing their sheen; its tusk points downwards like a dart, about to impale itself in the deck below. A sailor saunters by this lynching, adjusting his hat for the photographer's lens.

From another chain hangs another prize: a polar bear, caught up at its waist like a wet fur rug. It dangles snout-heavy, claws unsheathed as though only just yanked off the ice as it pawed at the water for fish. Behind it, the ship's laundry flutters in the breeze. A third photograph, almost unbearably sad, shows a young bear still clinging to its mother's dead body. Destined for life in a zoo, cubs were brought back in barrels barred at the top. Adults were chained to the mast like dogs. Sailors feared them more than whales: Horatio Nelson, who sailed to the Arctic on the unpropitiously named HMS *Carcass* in 1773, nearly died when he tried to kill a polar bear as a prize for his father.

A nearby canvas dramatizes just such a scene. Painted in 1829 by William John Huggins – later maritime artist to William IV, the Sailor King – it is entitled *Harmony*, after the main ship in the picture, although that is hardly an appropriate description. With an unerring eye for detail, Huggins has sought to record every activity carried out by the whaling fleet in the north. Presided over by a distant iceberg that erupts like a frozen flame from the sea, the picture presents an icy Eden under assault. In one corner bobs a baby-eyed walrus, plaintively addressing the viewer while three narwhals flee, tusks tilted high. A sailor stands over a seal, raising his club. The beast backs off towards the edge of the floe, silently barking. In the mid-distance, a bowhead brandishes its broad black flukes. Stuck in its back are two harpoons, and there's fire in the chimney. Birds scatter into the air.

Sailing through this bloodlust is the bringer of destruction, the *Harmony*, a barque of nearly three hundred tons. Around her masts are tied two pairs of jaw bones, trophies announcing a successful voyage in their own triumphal arch. Higher up hangs a garland, a circlet wound round with ribbons given by wives and lovers and tied aloft on May Day eve by the youngest married man. A relic of a medieval rite – attended with 'grotesque dances and other amusements' by men in strange costumes – it stayed in place until the ship reached home, when young cadets would race up the rigging to claim possession of the now weather-beaten wreath.

Below this Brueghelian spectacle, with its fleeting figures and vessels top-heavy with blubber and bones, the guilty parties are named: *Harmony* of Hull, *Margaret* of London, *Eliza Swan* of Hull; *Industry* of London; workaday ships, doing their job. We may look upon such scenes with horror, but a nineteenth-century Huller saw a vision of plenitude to be rendered in barrel-loads and marked by a whale's tail stamped in the captain's log. Such bloody acts – the plunging flukes, the spray the sailor felt on his face and the guts that spilt on deck – represented security from beggarly poverty, only ever a footstep away.

By 1822 Hull was England's most successful whaling port. One-third of the British whaling fleet sailed from there – thirty-three ships in 1830. Contemporary directories list more oil merchants than eating houses in the port, while maps show 'Greenland Yards' on the river banks where whale oil was processed, along with manufactories where whalebone was turned into 'SIEVES and RIDDLES of every description, NETS . . . for folding Sheep . . .' and 'STUFFING, for Chair and Sofa Bottoms . . . preferable to Curled Hair'. They have long since vanished, but other reminders of the industry survive in the city's museum. A sperm whale's deformed jaw hangs on the wall; a giant vertebra once used as a butcher's block stands on the floor; and ranged in a rack like billiard cues are ivory tusks which once formed a four-poster bed for some northern worthy –

> In old Norse times, the thrones of the sea-loving Danish kings were fabricated, saith tradition, of the tusks of the narwhale.
>
> The Pipe, *Moby-Dick*

– while in the middle of the room like a tobacconist's kiosk, a dimly lit case displays a row of glass-stoppered bottles.

Whale meat extract; an oily substance rich in protein
used in margarine manufacture etc.
Whalemeal prepared from powdered whale meat.
Used for animal feedstuffs.
Whale liver oil; a source of vitamin A.
Sperm oil; partially solidified. When refined it is
used for lubricating light industry.

Arthur Credland, the well-informed curator – himself a zoologist, and a man who has eaten both whale and seal – opens the cabinet and hands me a phial. The glass still feels oily as I tilt the amber liquid to and fro, slightly scented. Pure and transparent, this is what it all boils down to, a whale in a bottle. Now I recognize the sound as it winds around the room: the song of a whale, lamenting its long-dead cousins.

Leaving the city behind, suburban Hull gives way to the flat fields of the Holderness Plain. This is a levelled, alluvial landscape. Its coastline might look as though it is shouldering the storms, but this is where England is falling into the sea, at a rate of yards, rather than feet, every year.

Turning off one of the B roads that run inland but seemingly to nowhere, a rail-straight drive leads to the door of Burton Constable Hall. The Constables have lived in this elegant house with its red-brick towers and battlements since the sixteenth century, conserving their Catholicism in their private chapel while their land was eaten away by the waves. This far from London, no one really cared that the Popish faith was kept in the wilds of Yorkshire.

Out of season, the ticket office-cum-tearoom is empty. The woman behind the counter looks relieved. 'I thought for a moment you were wanting to look round the house.'

I set off in the failing light, only to be told by a passing groundsman that the bones I'm looking for were removed years ago. He tells me, hesitantly, 'I can show you some vertebræ.'

From the back of a shed filled with farm vehicles, his colleague emerges, rag in hand. 'He wants to know about the whale, Dave,' says my reluctant guide.

Taking a pencil stub from his pocket, Dave sketches out a shape on the claw of his excavator, outlining what looks like a giant fish bone. He describes how the skeleton once stood in the field beyond, articulated to mimic the animal in the water, held up by iron struts and bolted to a frame. The supports rusted away long ago; some Boy Scouts camping in the grounds had even attempted to make a fire with the remains.

But the bones had since been rescued. In the gloom of the outhouse, Dave pulls at a piece of sacking with the dramatic flourish of a pathologist drawing back a shroud. Underneath lies a great grey bone, eroded by decades of exposure to the weather; more a gigantic piece of coral than the skull of a cetacean.

'This is the only whale from *Moby-Dick* to exist,' he declares. The reality is as incongruous as his claim, all the more so for lying next to a disused caravan. This crumbling lump of calcium once held the animal's huge brain, controlling its sinewy muscles and the broad flukes and fins; listening to the watery world and watching it through sentient eyes; issuing mysterious clicks from its mountainous head.

In other outhouses lay the rest of the animal, scattered relics awaiting resurrection. Their bony diaspora was a measure of this martyred whale, ready to be reassembled to satisfy its modern pilgrims: scoured backbones the size of tractor wheels; pitted ribs resembling mammoth tusks dug out of Siberian permafrost; hulking masses of decaying calcium like debarked trees.

I walk to the end of an avenue of oaks, where a funereal urn is set on a crumbling plinth. In the tussocky grass to one side is an empty space. Bits of brick still lie in the turf, remnants of the foundations that once bore up the whale on iron waves. And as jackdaws caw in the darkening sky, I imagine the leviathan's bones luminous in the twilight. Could this really have been the whale of which Ishmael spoke, washed up in a muddy field in Yorkshire?

In April 1825 a dead whale was seen off Holderness, close to Burton Constable. There was nothing particularly unusual in that; such animals often wash up on this coastline, one of England's most desolate shores, where the North Sea eats away at the boulder clay leaving entire villages to crumble into the waves, and where fossilized forests lie under the surf. But this was a huge specimen, and as it floated out at sea, fishermen steered clear of the whale for fear it might damage their boat. Soon enough the tide did its job, and on the afternoon of Thursday, 28 April, the carcase was cast onto Tunstall beach. There, below the low, soft, chocolate-coloured cliffs, which turn the water a dolorous reddish brown, it was stranded like an enormous flounder.

The next day, Reverend Christopher Sykes, a keen amateur scientist, arrived to record the animal's vital statistics. By Sunday, a crowd of one thousand souls were drawn to witness this fabulous beast. Like their Dutch cousins across the sea two centuries before, they were amazed at what they saw: a bull sperm whale, fifty-eight feet long – although this was not the shiny black monster they might have expected. Thrown out of the sea, its proud jaw was dislocated and most of its paper-thin skin had peeled, revealing a strange layer of 'fur' between it and the blubber – as if the whale were in disguise all along. Slumped on the cobbles, the putty-coloured carcase had already begun to decay, a process hastened by onlookers who hacked about the body, pulling out the long, thick tendons and using horses and ropes to tear out the throat.

All Holderness was alerted to this deputation from the depths, as twenty-six-year-old Sara Stickney reported. 'You will doubtless have heard of the monster washed up on this shore – the bustle it occasioned in the neighbourhood was marvellous.' The village was 'more gay than sweet,' she confessed, 'the whale becoming every day more putrid – it was a loathsome thing at best. I never could tolerate the sight of an inanimate mass of flesh in any shape.'

The whale was soon rendered unrecognizable. Men cut into its huge head; the liquid looked like olive oil, but soon began to coalesce. At eighty degrees Fahrenheit, it was nearly thirty degrees warmer than the outside air, although the investigators could not determine whether this was a result of animal heat or of 'putrescent fermentation'. As it came apart, the wonders of the beast were examined. Its blubber, once tried out, would fetch £500; the case yielded eighteen gallons of spermaceti; and the meat would have fed a few families for weeks (one Hull recipe claimed that the animal's skin made a tasty dish, with a mushroom flavour). However, this foundling had a scientific value greater than its commercial or culinary worth; and, accordingly, Dr James Alderson was appointed to perform a post-mortem.

Son of a well-known Yorkshire physician, Alderson was a Fellow of Pembroke College, Cambridge, and of the Cambridge

Philosophical Society; his encounter with the whale was a chance not only to address the contradictory evidence about these creatures, but also to further his academic status. 'Nothing can be more contrasted than the view of the animal perfect, and its skeleton,' Alderson told his fellow society members, having had the extant parts brought to his laboratory in Hull. 'The enormous and preposterous matter upon its cheeks and jowl bearing no proportion to that of any other animal whatever, when compared with the bones of the head.'

The sheer logistics of examining this mountain of blubber presented Alderson with his greatest challenge – even as he received it in instalments. The whale's eyes had already been removed, being small and oddly shaped, 'in the form of a truncated cone', although they presented their own exquisite beauty. Alderson described the iris as 'bluish-brown; very dark; the pupil . . . transverse, as in ruminating animals'; while 'the tapetum presented a very beautiful appearance . . . its color was a green, formed by an admixture of blue and yellow; with a slight predominance of blue . . . speckled with lighter colored spots throughout.' That the doctor could discern such pulchritude in this mass of decaying flesh was a measure of the animal's fascination. Its parts presented themselves as if to say, Look how beautiful I was when I was alive, when I scooped up squid from untold depths, when I dealt death myself.

Embedded in its lance-like lower jaw were forty-seven teeth, scarred with its adventures in the abyss. Alderson observed that the penis 'protruded about 1½ feet from the body, and was surrounded by a shaggy process of the cuticle. The urethra admitted the point of the finger.' Fingering the whale was ever a common abuse. The three-foot-long heart was preserved in formaldehyde, and later presented to the Yorkshire Philosophical Society for their further ruminations.

Alderson was frustrated by the treatment of his specimen: 'indeed, the viscera were so quickly removed, with a view to clearing the bones of the animal, that it was impossible to examine every organ.' For all his delving, the doctor could find no cause of death, despite the presence of a five-inch section of a swordfish spear buried in its back, 'enveloped in the adipose cellular membrane', as well as another 'fistulous-like opening in the cutis' apparently made by a harpoon. Sperm whales were known to carry foreign objects in their flesh, like shrapnel in a soldier's war wound, and Thomas Beale recorded swordfish attacks on whales. One animal was found with an entire swordfish blade in its dorsal ridge, the result of a violent collision during which the weapon had slid clean into the whale, snapping off at the base; when the scar healed, the sword remained embedded in blubber, a fishy Excalibur. Similarly, Ishmael says that harpoons could lodge in a whale, one entering 'nigh the tail, and, like a restless needle sojourning in the body of a man, travelled full forty feet, and at last was found imbedded in the hump'.

Speared and nailed, whales carried not only the scars of their martyrdom, but the instruments of torture themselves. Yet even as this unfortunate, war-weary creature was washed ashore, its destiny was assured: for as Ishmael notes, from that moment on, the whale became the property of the Lord Paramount of the Seigniory of Holderness, the Squire of Burton Constable Hall.

Almost anywhere else on the English coast, the Tunstall whale would have belonged to the Crown, a royal fish; but here, on what amounted to a personal fiefdom from the cliffs of Flamborough Head to the filigree finger of Spurn Point, the Lord Paramount exercised that right. One of the first men on the scene had been the Constables' own steward, Richard Iveson, come to lay claim to this odoriferous prize. Iveson had measured and sketched the body as it lay on the beach, as if to reinforce his master's right. His drawing was subsequently reproduced as an engraving, the accuracy of which would not have pleased Ishmael, resembling as it does a giant tadpole, with an out-of-scale surveyor – Iveson himself – striding across its head.

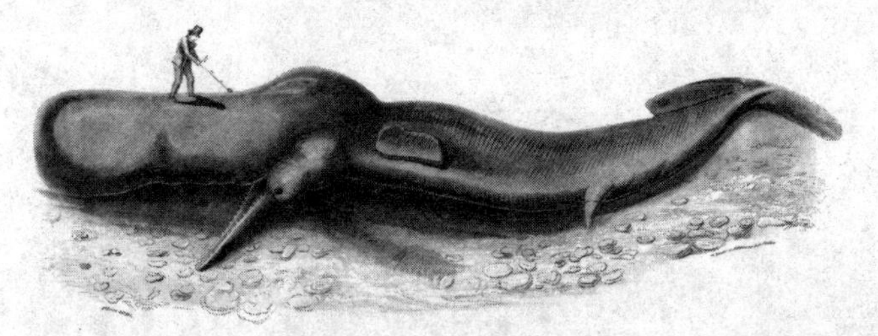

More accurate were the illustrations made by Alderson's brother, Christopher, and included in the doctor's *An Account of the S. Whale Cast on Shore at Tunstall, 1825*, a copy of which, suitably bound in red morocco and stamped with gilt flourishes, was subsequently presented to the Lord Paramount. This portrait of the whale is distinctly romantic, every curve and undulation lovingly shaded like the Rokeby Venus – an impression reinforced by the animal's oddly waisted form and feminine hips, despite the exposed member close to its lackadaisical tail. It is seen from front and rear, from every enticing angle, while yachts flutter in the distance, lending a lyrical air. More pathologically, a second illustration displays its jaw and skull in close-up, as well as a study of the eye, sliced open to display its beauty. A third shows a squid beak, one of a bucketful found in the belly of the beast.

Although a stranded whale might represent a valuable contribution to the Constables' coffers – in 1790 a whale found at Little Humber yielded 85 gallons of oil at 9d per gallon – the accounts of previous stewards show that the costs often exceeded the profits.

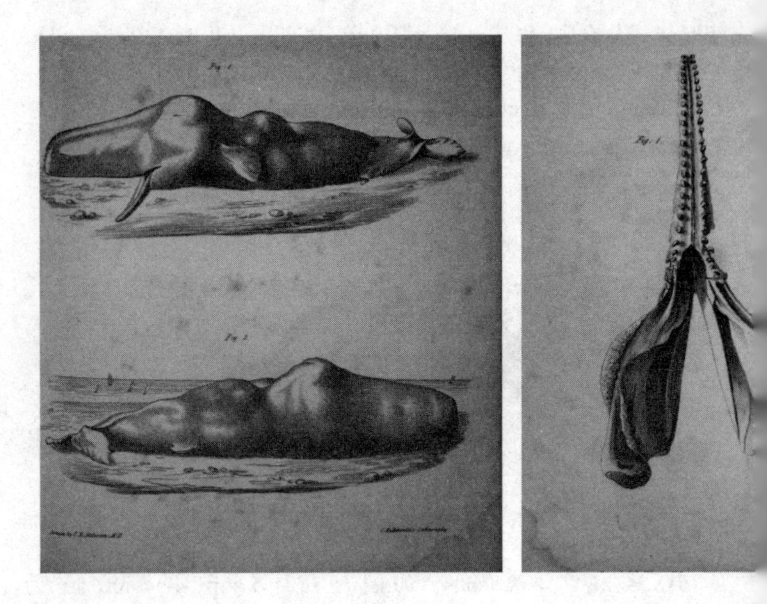

Minutes of Escheats, Deodands, Royal Fishes, Wrecks, &c.

John Raines, Steward to William Constable.

Jan. 30th 1749. A Sperma Ceti Whale was thrown on the shore at Spurn Point – Mr Constable sold it to Mr David Bridges of Hull for £90.

Sept. 13th 1750. A Whale 33 yards long was thrown on shore upon Spurn Point. Mr George Thompson cut it up for Mr Constable – Mr Thompson's charge for the Exps of cutting up amounted to £7 more than the Whale sold for.

Nov. 7th 1758. A Grampus came on shore at Marfleet – Mr Constable sold it to Mr Hamilton Merchant of Hull for £5. 10.s.

Nov. 9th 1782. A Whale 17 yards long came on shore at East Newton – It was sold for one Guinea & an half, being much damaged, & in a state of putrefaction.

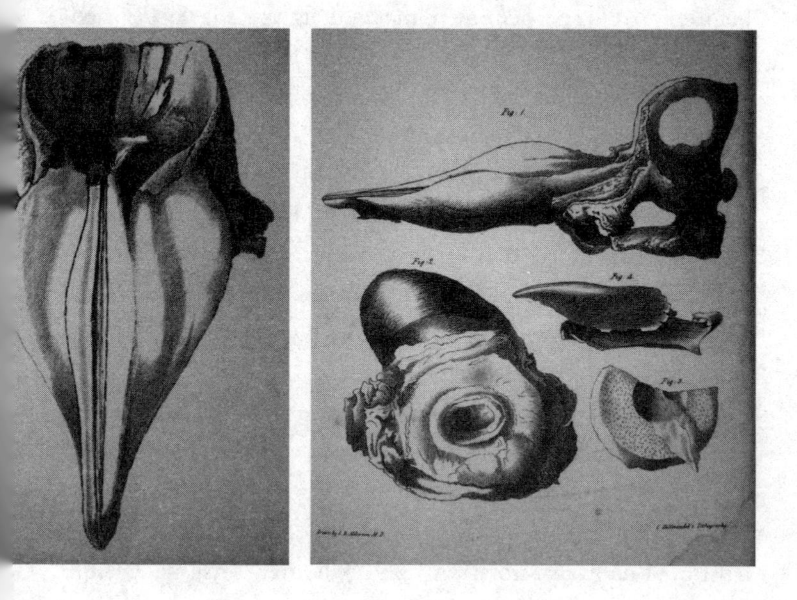

> *Jul. 14th 1788.* A Whale 36 Feet long, came on shore at Spurn Point, opposite the Lights upon the Humber Side. Mr Pattinson the Baliff sold it to Mr De Poyster of Hull for £7. 7.s – But it proved good for nothing, having died of poverty . . .

Whatever its financial benefits, the Tunstall whale was destined for a different fate. Sixty years earlier, the Bishop of Durham had laid claim to another of the spermaceti tribe thrown up on the north-east coast – a fifty-foot 'Sea Monster' still alive when it was beached at Seaton in 1766, where its 'calls of distress as it touched ground could be heard for several miles', and whose skeleton was later displayed in the undercroft of the cathedral; an image that reminds me of William Walker, the Victorian diver who was sent into the flooded foundations of Winchester Cathedral to shore up its medieval timbers. So too the Yorkshire whale was to be preserved for perpetuity. To that end, its remains were buried in a series of pits, and there they were left to rot.

The new owner of Burton Constable Hall, Sir Thomas Aston Clifford Constable, second baronet, was just eighteen years old in 1825, and whales were far from his mind. He was more concerned with the business of spending the substantial legacy he had just inherited. Two years later, Sir Thomas married Marianne, youngest daughter of Charles Chichester, and his Yorkshire pile lay empty while its lord lived in Staffordshire, closer to London and its amusements.

Dry bones proved unequal to such diversions. While its owner enjoyed the fruits of his fortune, the whale's skeleton, now picked clean and disinterred, languished 'in a very neglectful condition, being laid in an irregular heap, in the middle of a field', as one frustrated naturalist noted in 1829. 'Whether it has since been put together and taken care of, I have not heard.' Seven years later, little progress had been made in the matter. The geologist John Phillips found the bones in a barn, save those of the tail, which unaccountably hung in a tree. Then, in 1836 – when Sir Thomas finally deigned to move to his ancestral home – Edward Wallis, surgeon, anatomist and astronomer, was engaged to articulate the whale: to give it life after death.

In the second quarter of the nineteenth century, whales were suddenly *à la mode.* Popular interest in science and natural history met sensationalism and showmanship, and whales were exhibited around Europe and America, preserved or in skeletal form. In March 1809 'the curious were gratified' by a seventy-six-foot 'stupendous monster of the deep' shown on a barge moored on the Thames between Blackfriars and London Bridge. The whale was claimed to be a year old, and 'pronounced by judges to be the *Balena Boops,* or pike-headed species' – a confusion of the former

Latin name for the humpback and another common name for a minke, neither of which reach such lengths. 'But the prudence of bringing into the centre of a popular city, for the mere gratification of gazers, a monster of such bulk, in *a state of putrefaction* is quite another consideration,' queried *The Times*. 'In all events, those who visit the whale will do well to use the expedient of holding to their mouths and noses handkerchiefs well moistened with strong vinegar, to guard against inhaling the putrid effluvia it emits, than which few things can be more noxious to health, and even life.'

Other showmen had the acumen to make their displays fit for more refined tastes. In 1827 a blue whale taken off Ostend was reduced to its skeleton and toured from Ghent to Brussels, Rotterdam and Berlin before arriving, four years later, in London, housed in a custom-built wooden pavilion – 'a wondrous lengthy booth' – at Charing Cross, close to where Melville would stay. *The Times* claimed – virtually in the voice of the fairground barker – that at ninety-five feet, the whale was 'of larger dimensions than any that is known ever before to have fallen into the possession of man'. Visitors paid a shilling to enter what one doggerel called 'a tomb/ A sort of bed-crib, sleeping room/ For what they call – a Whale'. The hut was stocked with volumes of Lacépède's *Natural History of Whales*, and customers could quaff wine while sitting in the animal's ribcage, an 'unwonted saloon'. They were, however, not treated to the twenty-four-piece orchestra that performed within the whale during its European sojourn.

Whales were the sensation of the age. A few years later, England produced its own regal specimen, claimed to be even bigger –

As a man of fashion and taste himself, Sir Thomas now saw fit to have his own whale put on display. Its spine was duly riven with an iron rod, its ribs hinged with stirrup-like irons, and long bolts were driven through its skull. Artificially jointed, the skeleton was

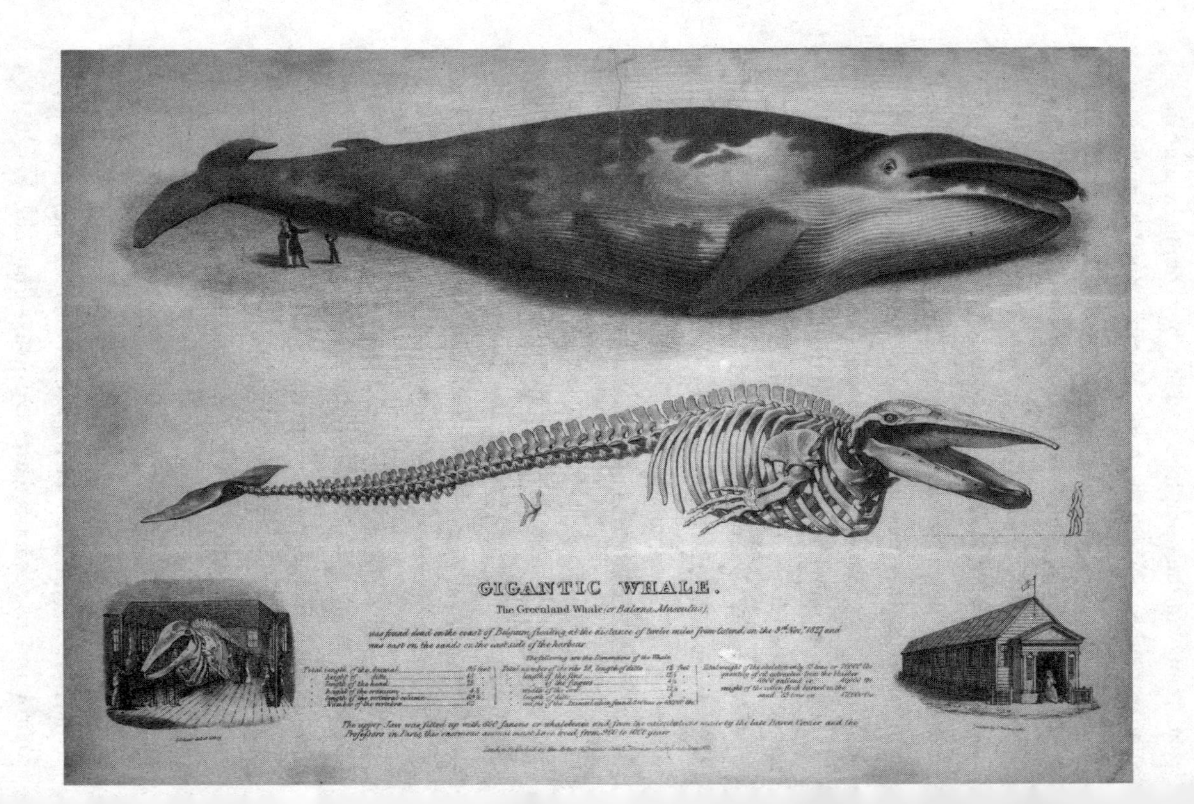
GIGANTIC WHALE.
The Greenland Whale (or Balena Musculus).

set to swim along an avenue of trees which became known as the Whale Belt. It was here that Thomas Beale – the foremost authority on sperm whales – came to pay his respects. Alerted by Mr Pearsall, curator of the museum of the Hull Literary and Philosophical Society, Beale sought an audience with the Yorkshire specimen; with his arrival in the East Riding, the whale would be made immortal.

Unlike many scientists who pronounced upon the subject, Beale had actually seen living sperm whales. As a young man, he had studied medicine at Aldersgate from 1827 to 1829, remaining as an assistant in the school's dissecting room, then as curator before moving to the London Hospital on Commercial Street. But in 1830, at the age of twenty-two, Beale left the grimy streets of the East End to sail on the whale-ship, the *Kent*, captained by William Lawton and owned by Thomas Sturge.

Beale's journey took him down the coast of South America to Cape Horn, then across the Pacific to Hawaii and on to the Kamchatka Peninsula – almost as far from England as it was possible to be. During his travels he watched whales being hunted, making extensive notes about their behaviour and physiology, gathering scientific information in a manner that echoed the work of Charles Darwin, whose own voyage on the *Beagle* was under way even as Beale reached the South Seas.

While Beale was fascinated by the life beyond his ship, he was appalled by the oppression on board it. 'When I saw thirty-two good, industrious, and harmless, though brave men, abused and browbeaten to a most shameful extent, by a mean and contemptible tyrant . . . I turned from the scene with horror, and plainly intimated that I could no longer endure the sight.' At midnight on 1 June 1832, at the Bonin Islands, Beale jumped ship, joining another Sturge whaler, the *Sarah and Elizabeth*, under whose more temperate captain – who happened to be the gallant William Swain, later master of the *Christopher Mitchell* and himself to lose his life to a whale – he sailed home, having travelled fifty thousand miles.

Like Thoreau, Beale's experience of whales left him amazed at the lack of knowledge about them. 'It is a matter of great astonishment that the consideration of the habits of so interesting and in a commercial point of view of so important an animal, should have been so entirely neglected,' he wrote. 'In fact, till the appearance of Mr Huggins' admirable print, few . . . had the most distant idea even of the external form of this animal; and of its manners and habits, people in general seem to know as little as if its capture had never given employment to British capital, or encouragement to the daring courage of our hardy seamen.' Beale referred to William John Huggins's *South Sea Whale Fishery* – an image so enduring that it was still being used as the basis for a *New Yorker* cartoon in the first decade of the twenty-first century.

Back at his Bedford Square home, Beale set out to correct the cetological lacunæ. A year later, he presented his paper on *Physeter macrocephalus* to the Eclectic Society of London, which awarded him their Silver Medal for his efforts. Having published his text as an elegant, illustrated booklet in 1835, the surgeon spent the next four years working on an expanded version. *The Natural History of the Sperm Whale*, published in 1839, was a wide-ranging and eclectic work, part scientific study, part adventure story. Its frontispiece (and this chapter's) is an all-action scene showing sperm whales in an ocean they have whipped up into a freeze-framed, foam-flecked frenzy as they toss boats out of the water and send harpoons and humans flying into the air.

Equally evocative are Beale's chapter headings, succinct summaries of his experiences on far-flung oceans and among exotic peoples.

> *the Author is robbed – Sea-Lion fight – Music of the Birds – Shocking Diseases – instances of religious tyranny – an Apprentice drowned – narrow Escape – intense heat – we kill a female Whale – a Dandy Savage – a Necromancer – Tyranny of our Captain – Six Men flogged – I leave the "Kent" at midnight – see immense numbers of large Whales – a young man is bled – a Bolabola Girl's Eyes – we are invaded by thirty Women – three Men washed from the Jib-boom – crossing the Line the sixth time – Reflections on seeing our Native Land – stern Disease has been raging during our absence – we approach Home with faltering steps – the old House – my emotion and fate trade.*

Beale's narrative – its retelling of myths a presentiment of Sir James Frazier's anthropo-religious *The Golden Bough*, its human exploits redolent of a picaresque novel – provided a framework for Melville; an articulation for his own whale. If *Moby-Dick* owed its metaphysics to Nathaniel Hawthorne, then it owed its facts to Thomas Beale. Entire passages in Melville's book are filched directly – one might say almost brazenly – from Beale's. *The Natural History of the Sperm Whale* was the archetype for *Moby-Dick*; not only in its cetological details, but in its other preoccupations, too.

Beale was intrigued by the whale's role in the human chain of consumption. It was as if, in the spirit of emancipation, he saw the whale as enslaved – a notion underlined by the dedication of his book 'to Thomas Sturge, Esq, of Newington Butts'.

> As the trusty friend of MACAULAY, you fought the battle of the Negro . . . and it was not until the enemies of the dark human race began their precipitate retreat, that the wavering friends . . . flocked around the banner you had helped raise . . . And now that the Negro is free . . . I have no doubt . . . that your greatest reward is in your own feelings, independently of worldly praise.

Thomas Sturge was scion of an old Quaker family; his kinsman was the even more famous abolitionist, Joseph Sturge. He owned the two ships on which Beale had travelled, and, like his friend Elhanan Bicknell, ran his whaling company from the New Kent Road, near the Elephant and Castle in south London, a decidedly unoceanic address. (He also benefited from strandings. When the sperm whale of which Buckland wrote beached at Whitstable in the winter of 1829, to the accompaniment of terrible bellows and groans – and was attacked by man with an axe for its pains – Sturge paid sixty shillings for its blubber.)

There was a certain distance between these refined men and their noisome business. Bicknell – who held the monopoly on the British sperm whale fishery in the Pacific – was a well-known patron of the arts who commissioned Huggins to paint his whale-ships –

works that would in turn inspire J.M.W. Turner, another beneficiary of Bicknell's patronage. In this complex web of connections, whales linked writers, artists, scientists and businessmen in a manner that reflected the reach of the British Empire and the very size of the animals themselves. Whales lent a romantic focus to their gruesome industry. Indeed, Turner, the greatest artist of the age, realized that vision in paint, just as Melville attempted it in words.

In 1845 and 1846 Turner exhibited four scenes of whaling at the Royal Academy, along with a catalogue attribution: '*Whalers. Vide Beale's Voyage p.175*'. They portray the heroic hunt for the whale in luminous, almost abstract forms; the whales themselves are the merest, ghostly suggestions. It is likely that Melville had heard of these famous pictures on his visit to London. Certainly, back in New York, having bought his copy of Beale's work – for three dollars and thirty-eight cents – he in turn wrote on its title page, 'Turner's pictures of whalers were suggested by this book'.

Melville's passion for Turner almost rivalled that of the artist's champion, John Ruskin. (Critics themselves drew the comparisons: one reviewer of *White-Jacket* declared, 'Mr Melville stands as far apart from any past or present marine painter in pen and ink as Turner does from the magnificent artist vilipended by Mr Ruskin for Turner's sake – Vandervelde.') Turner appealed deeply to Melville's sense of the romantic. In his book, *Modern Painters* – which Melville read before his trip to England – Ruskin described how Turner had himself tied to a ship's mast to paint his *Snowstorm at Sea.* Perhaps the painter had more than a little of Ahab in him.

The influence of Turner's sublime vistas, numinous with storms and shadows, emerges in *Moby-Dick* from the first. When Ishmael arrives at the Spouter Inn, he sees 'a long, limber, portentous, black mass of something . . . floating in a nameless yeast'. Through the gloom, he makes out a whale launching itself over a storm-tossed ship, seemingly about to impale itself on its masts. 'A boggy, soggy, squitchy picture truly, enough to drive a nervous man distracted,' Ishmael allowed, with 'a sort of indefinite, half-attained, unimaginable sublimity about it that fairly froze you to it, till you involuntarily took an oath with yourself to find out what that marvellous painting meant'. It was a dream version of Huggins's graphic scenes; a fantastical Turner, seen through Ishmael's ostensibly amateur eyes.

From these colourful scientists and eccentric artists, and from his own visit to Liverpool and London, Melville's enterprise acquired an English anchor. Their personæ, as much as their efforts, were integral to the intricate tapestry of cross-references and diversive threads from which *Moby-Dick* was woven. Above all, it was Beale who supplied Ishmael's cetology, and who sought to correct those erroneous pictures of whales, applying true science and firsthand experience to the natural history of the sperm whale. He criticized the respected French naturalist Baron Cuvier, for instance, for claiming that the whale struck fear into 'all the inhabitants of the deep, even to those which are the most dangerous to others; such as the phocæ, the balænopteræ, the dolphin, and the shark. So terrified are all these animals at the sight of the cachalot, that they hurry to conceal themselves from him in the sands or mud, and often in the precipitancy of their flight, dash themselves against the rock with such violence as to cause instantaneous death.'

To Beale – as to anyone who had seen sperm whales in the wild – this was so much hogwash. 'For not only does the sperm whale in reality happen to be a most timid and inoffensive animal . . . readily endeavouring to escape from the slightest thing which bears an unusual appearance, but he is also quite incapable of being guilty of the acts of which he is so strongly accused.'

Beale comprehensively addressed every aspect of the whale, point by point, fin by fluke. Yet no matter how many facts and figures, how many observations he assembled, no matter what physiological detail – from the function of its stomach to how much blubber its carcase might yield; from its 'favourite places of resort' to the 'rise and progress of the Sperm Whale Fishery' – his quarry remained elusive. Only by laying his hands on the very bones of the animal could the surgeon make his final diagnosis; and even then, he might wonder at the reality of the beast he pursued.

The sperm whale had taken Beale halfway round the world. Now it summoned the surgeon to east Yorkshire, by no means an easy journey. Having made his way to Holderness, Beale was rewarded for his efforts by a spectacular sight: a skeleton key to the innermost secrets of *Physeter.* He may have seen the animal in life, but in its decay its true nature was revealed, and he was enthralled by what he saw. 'The description of the skeleton of the sperm whale at Burton-Constable, which I shall presently give, interests me exceedingly, principally on account of its being the only specimen of the kind in Europe or in the world.'

Practically falling over himself in his eagerness to get at the bones, Beale lost no time in making notes on 'this enormous and magnificent specimen of osseous framework'. His report extends for many pages: 'Extreme length of the skeleton 49 feet 7 inches' – the shrinkage being due to the creature's unboned flukes and blubber – 'extreme breadth of the chest 8 feet 8½ inches . . . The gigantic skull . . . forms more than a third of the whole length of the skeleton . . . The lower jaw is 16 feet 10 inches long . . . The spinal column consists of forty-four vertebræ . . . In the lower jaw there were 48 teeth.'

Beale's examination endowed the Tunstall whale with eternal life. This was the first accurate description of a sperm whale skeleton; it became the *ur*-whale, the whale by which all others would be measured. Seen through Melville's literary lens, these bones acquired a kind of poetic licence. They pervade *Moby-Dick.* Dave was right: the jumble of ribs and vertebrae he showed me in a Yorkshire outhouse were indeed the only physical relics of Melville's book; and they achieved their place in perpetuity via Beale's ground-breaking book. When his own copy of *The Natural History of the Sperm Whale* surfaced a century later, Melville's marginalia had been erased by an owner who had little idea that they were worth more than the volume itself. Enough marks remained to show that the book supplied the scaffolding for *Moby-Dick*'s construction; and that Melville specifically drew on Beale's notes on the Tunstall whale to create an elaborate conceit – one that fused his own visit to St Paul's

Cathedral with those travelling exhibitions of whale carcases and skeletons that had become so fashionable. The result was an arch architectural exercise in irony, a wry and witty metaphor for man's use of the whale.

> Sir Clifford's whale has been articulated throughout; so that, like a great chest of drawers you can open and shut him, in all his bony cavities – spread out his ribs like a gigantic fan – and swing all day upon his lower jaw. Locks are to be put upon some of his trap-doors and shutters; and a footman will show round future visitors with a bunch of keys at his side. Sir Clifford thinks of charging twopence for a peep at the whispering gallery in the spinal column; threepence to hear the echo in the hollow of his cerebellum; and sixpence for the unrivalled view from his forehead.

Yet in his gentle satire, Melville could not know that, only months before his visit to London, the author of this seminal work had died in that city, aged just forty-two. For ten years Beale had worked as medical assistant to the Royal Humane Society; he also joined the Institut d'Afrique, a Parisian organization pledged to the welfare of slaves, and spent the rest of his life as a poorly paid officer at the Stepney Poor House in the East End. There, while caring for his patients in the cholera epidemic of 1848–9 which had claimed 60,000 lives, Beale contracted that same 'stern Disease'. Within twenty-seven hours, this ultimately humane man was dead.

The narrow stone-flagged and dark-panelled corridor gives way to gracious Georgian rooms, all wrapped up for the winter. A cantilevered stairway turns creakingly on itself, without obvious means of support. It is early in the morning; the house is empty.

I open door after door, finding bedrooms filled with exquisite marquetry wardrobes, elegant chaises longues and beds covered in embroidered velvet. On one trunk lies a discarded military frock coat, as if its owner had just stepped out of the room. At the other end of the landing stands a pair of mirrored double doors, and beyond them, the Long Gallery.

Once this was used for indoor recreation, for fencing or strolling in inclement weather. Now it is lined with bookcases and a plaster frieze in seventeenth-century style. It depicts a veritable menagerie of chimerical and transgendered beasts. One has a woman's torso and breasts, but a stallion's body and penis. Another shows a snarling, scaly Jacobean whale, fighting to free itself of its entablature; all teeth and flukes, it heads down the hall towards its time-honoured opponent – a giant squid splayed above the door, flanked by a curly-tailed mermaid.

This antique animation carries on, regardless of the silence of the room, orchestrated by its commissioner, William Constable, whose own portrait hangs below. He is clad in a Rousseauesque gown and turban, a man of Enlightenment tastes – that much is evident from the contents of his cabinet of curiosities, now housed in an anteroom at the end of the gallery. Like the Quakers, Constable was barred from high office by his faith; and just as they directed their energies into business – the business of killing whales – so the squire of Burton Constable was excused the expense of political service, and could spend his considerable fortune elsewhere.

Chemistry, astronomy, botany, zoology and ancient history all clamoured for Constable's attention: from ornate shells and polar bear skulls to casts of Roman and Greek coins kept in specially made cases. One cabinet contains early electrical equipment, an elaboration of hardwood wheels, brass cylinders and rubber belts producing sparks to be stored in glass Leyden jars, ready for a Frankenstein experiment. On another shelf lie relics of a true monster: the teeth of the Tunstall whale, arrayed as though newly pulled from a dragon's jaw.

John Raleigh Chichester-Constable, the current tenant of Burton Constable Hall, is a dapper man in tweeds, cravat and Geo. F. Trumper cologne. He recalls how, as a boy, more than seventy years ago, he would play in the whale's skeleton which then stood in the grounds, using it as a giant climbing frame. As heir to the Seigniory, Mr Chichester-Constable is still notified when any cetacean is thrown on this coast, and may dispose of it as he will. He once took a dead porpoise into Hull to have a pair of fashionable ankle boots made for his wife from its skin, only to be asked by the cobbler – who, he claimed, was a relative of Amy Johnson, the aviatrix – to take the carcase out of his shop before its smell drove his customers away.

As a young man, Mr Chichester-Constable was also an amateur pilot, landing his private plane on the long narrow field next to the Whale Belt, while the whale looked on, in an ever more dilapidated state. It endured the decades, exposed to the pouring rain, the freezing frost, the blanching sun, neglected in the nettles and long grass, awaiting the day it would be revived. On a late summer's day in 1996, the bones were exhumed by Michael Boyd, zoologist and historian. Like Melville before him, Boyd was assisted in his task by Beale's description in *The Natural History of the Sperm Whale*; by referring to his nineteenth-century predecessor, Boyd was able to salvage most of the skeleton.

It was a hot afternoon, and exhausted by his efforts as he worked in his shirt sleeves and vest, Boyd felt as Ahab had felt about 'thou damned whale'. Although the Victorian articulation had corroded, he still had to saw through thick iron bars before the great ribs and vertebræ appeared, remarkably preserved, not unlike the ichthyosaurs he had excavated from the strata of nearby Robin Hood's Bay. Slowly, the whale emerged, bit by bit, bone by bone. The skull was still riven by rusty bolts as though it had undergone some ancient and rudimentary cranial surgery. And when the jaw bone was uncovered – split in half like a giant wishbone – an unerupted tooth was found in it, as if the whale had reverted to infancy in its interment.

Now the result has been brought into the great hall, where it lies on the floor, overlooked by ancestral portraits and narwhal tusks, like a hunted tiger laid out for its master's delectation. In a house filled with strange beasts – dead-eyed impala impaled on the walls, and silver-gilt Chinese dragons crawling up the window frames – the whale is an elegant whimsy to greet modern visitors. Yet its bones represent only a reduction of the animal. Alive, it would not have fitted into this huge chamber. Its forehead would have nudged the doors, and its flukes would have squashed against the landscapes hanging on the far wall, like a salmon squeezed into a goldfish bowl.

X

The Whiteness of the Whale

> Deathful, desolate dominions those; bleak and wild the ocean, beating at that barrier's base, hovering 'twixt freezing and foaming; and freighted with navies of ice-bergs . . . White bears howl as they drift from their cubs; and the grinding islands crush the skulls of the peering seals.
>
> Herman Melville, *Mardi*

Driving north from Burton Constable, the years fall away with the coastal road, running through familiar names: Bridlington, Filey and Scarborough, childhood memories of amusement arcades and fish and chips, and the burnt-sugar smell of candy floss, and pale green gas-mantle lamps hissing into the night, as fragile as the moths that flutter round them while my mother made tea in our caravan.

If the past is a contraction of what has passed, then the future exists only if we imagine it. These resorts recede into memory, and safe fields yield to wild moorland, wide expanses of nothingness book-ended by impenetrable plantations of black conifers. The car radio turns into white noise as we pass the giant white golf balls of the Fylingdales listening station. Then the road descends to Whitby, another half-hidden place, with its ancient red roofs and its steep streets and snickleways coursing down to its horseshoe harbour.

Here, among these narrow terraces, lived my great-grandfather, Patrick James Moore; a Catholic, too, albeit born to rather less propitious circumstances than the tenants of Burton Constable. The son of a Dublin blacksmith, he had joined the general exodus from Ireland, passing through the same Liverpool docks that Melville had explored; one of Melville's shipmates on the *St Lawrence* was an Irishman named Thomas Moore. By 1882 Patrick Moore had arrived in Whitby, with his wife Sarah, a house-maid from Faversham who, six months after they were married, gave birth to their first child, Rose Margaret. Perhaps that's why they lived in a poor part of town at Grove Street, close to Scoresby Terrace; although at the end of the lane were the riverside works where James Cook's ship, the *Endeavour*, was built.

It was there that my grandfather, Dennis, was born in 1885. He would grow up to be a tailor, making suits for J.B. Priestley and an overcoat for Winston Churchill, but by the time I knew him, at the end of his life, he had retired to Morecambe – known as Bradford-by-the-Sea – where he would die in a home facing the great expanse of the bay. I have only vague memories of his visits to us: a dapper, white-haired old man dressed in elegant dark suits. He always wore a watch and chain and, as my parents told me, had such a passion for reading that he would often miss his bus stop, so engrossed was he in his book. I was a young boy, and I had no idea that my grandfather had been born in a town that lived with the memory of whales.

Still less did I know that, around the same time as my young grandfather was playing in its streets, Bram Stoker was holidaying in Whitby, a stay that inspired his most famous work, the sensational story of *Dracula*. In it, Stoker's heroine Mina climbs the steps to the town's clifftop graveyard, where she meets an elderly man who had sailed to Greenland 'when Waterloo was fought', and who tells her 'about the whale-fishing in the old days'. This old sailor, nearly one hundred years old, is a relic of Whitby's past, and an industry carried out, not in the balmy seas of the Pacific, but in the freezing wastes of the Arctic: the wilderness at the top of the world.

In Edgar Allan Poe's only novel, *The Narrative of Arthur Gordon Pym of Nantucket*, published in 1838, a sixteen-year-old stowaway sails on a mutinous whale-ship out of New Bedford. After murder and shipwreck, Pym and his companions are forced 'to this last horrible extremity' – to dine on their young shipmate, Richard Parker. Poe's tale – which Melville must have read – was inspired by the fate of the *Essex*; it also had a strange reverberation forty years later, when the survivors of a shipwrecked yacht sailing from Southampton for Australia ate their own cabin boy. By remarkable coincidence, his name was also Richard Parker, and his memorial in the local churchyard, close to where I grew up, forever fascinated me with its ghoulish epitaph: *Though he slay me yet will I trust in him.*

But Poe's story has other resonances, too, in the lands and creatures that young Pym encounters on his subsequent adventures in the Antarctic, where he sees ice bears with blood-red eyes and murderous Indians with black teeth. Drawing on notes made by his friend, Jeremiah Reynolds, who had undertaken his ultimately disastrous expedition to the Antarctic in 1829 (Reynolds's crew mutinied on the way back, forcing him off the ship at Chile thus providing the setting for his own story of Mocha Dick), Poe presented his book as non-fiction. He even told friends he had been a whaler himself. Newspapers ran excerpts as factual accounts, convincing readers of a new and unknown land where the waters grew warmer rather than colder as one travelled towards the pole and where superstitious natives regarded anything white as taboo, fearful of 'the carcass of the *white* animal picked up at sea', and 'the shriek of the swift-flying, white, and gigantic birds which issued from the vapoury *white* curtain of the South'.

As its adventurers sail into the furthest unknown, they encounter a 'shrouded human figure, very far larger in its proportions than any dweller among men', its skin 'of the perfect whiteness of the snow'. This eerie otherworld, teetering between travelogue and science fiction, was the birthing-ground for Melville's monster. It is the source of the whiteness that appals Ishmael and on which he expands compendiously, if erratically, like a nineteenth-century search engine: from albino humans, 'more strangely hideous than the ugliest abortion', to 'the tall pale man' seen in the forests of the 'fairy tales of Central Europe . . . whose changeless pallor unrustlingly glided through the green of the groves'. Whiteness for Ishmael is as much the colour of evil as of good; it is an intimidating absence: 'Witness the white bear of the poles, and the white shark of the tropics; what but their smooth, flaky whiteness makes them the transcendent horrors they are?'

But that whiteness was also an invitation. Before the last of the wild lands were mapped, it was left to storytellers to fill in the fictional spaces on the map – from men such as Poe who had never travelled further than New England, to the mulatto boat-steerer Harry Hinton of *Nimrod of the Sea*, who imagined a shining wall of ice beyond which was an open sea, home of mermen and krakens with golden antennæ; a sanctuary where 'worried whales find peace, and grow in blubber on the crimson carpets of medusæ', safe from hunters who sought 'to harpoon and lance, to mangle, tear, and boil'.

THE SEA BEYOND THE SHINING WALL.

Such imaginings crept into what claimed to be reality, too. In a remarkable frontispiece created for Oliver Goldsmith's encyclopædic *Animated Nature* ('with numerous notes from the works of the most distinguished British and foreign naturalists'), first published in 1774, but subsequently reissued 'for the young and tender', as Ishmael observes, the artist assembled

the known denizens of the frozen world, drawing freely on William Scoresby's *An Account of the Arctic Regions*, to the extent that its decorously beached narwhal and bucking whale busy throwing its assailants in the air are direct imitations of Scoresby's pictures.

Yet among these seals and sea lions – themselves assailed by a ferocious polar bear – sea eagles and auks and walruses, a sea serpent swims blithely through the scene. It is perfectly at home among water spouts and icebergs, all the while observed by another narwhal, as if there were nothing extraordinary about its appearance at all; as if its existence were, by virtue of the many reports of its dalliances in other seas, established as a biological fact, to be represented alongside the other fauna of the polar ocean – even though further inspection reveals that this too is a crib of a creature, copied from the maned monster depicted in Pontippidan's *Natural History of Norway*.

The myth and romance of the Arctic was implicit in its alternative names: the Barren Grounds, Ultima Thule, the North Pole. As white as it was, this was one of the world's dark places, spending six months in perpetual night, a land so inhospitable it might as well be another planet altogether. Its axial emptiness, both on the page and in the mind, made it a site of sublime extremes. Its virginal whiteness meant death for any living thing unaccustomed to it; yet its temperatures produced the most abundant seas on earth. Delicate snow crystals could freeze a man's blood; but they also preserved an icy paradise ruled over by the land's greatest predators, whose fur looked white but was in fact translucent and coal-black beneath. Meanwhile, in its limpid waters swam creatures stranger than any invented by Poe.

> And of all these things the Albino whale was the symbol. Wonder ye then at the fiery hunt?

The Arctic whales – bowheads, beluga and narwhals – are the most tantalizing of all cetaceans. Rising and falling with the changing seasons of ice, they are barometers of an invisible world, spectrally floating within their bounded sea, locked into its cycle. They are philopatrous animals, loyal to the site of their birth, and the only whales to live in the Arctic throughout the year. One hundred thousand belugas swim in polar seas; the geographical remoteness of the less populous bowheads and their outriders, the narwhals, is such that they are seldom seen.

The beluga and the narwhal are a family unto themselves, the only two species of the *Monodontidæ*. Belugas, or belukhas, *Delphinapterus leuca*, owe their common name to the Russian for white, *belyy*. Their whiteness is not that of an albino as Moby Dick was supposed to be, an animal made unearthly by being devoid of colour; rather, they are born grey, and only achieve pure white in late adulthood, becoming sinless with old age. Their malleable melons (which one observer describes as feeling like warm lard) and their articulated necks allow the beluga to change the shape of their heads, holding them at right-angles and lending them a quizzical, human expression. Sailors called them canaries of the sea on account of their songs – and William Scoresby depicted his white whale perched on rocks like a seal basking in the sun – but to me they look like labradors, snow puppies in search of a master.

BELUGA or WHITE WHALE.

The narwhal, too, shares the beluga's sad beauty, a mortality suggested by its name – from the Old Norse, *nar* and *hvalr*, meaning 'corpse whale', because its smudges resemble the livid blemishes on a dead body. (It is not the only cetacean with such morbid overtones: the Latin name of another Arctic visitor, the killer whale – or, more correctly, whale-killer, *Orcinus orca*, has its root in *orcus*, meaning 'belonging to the kingdom of the dead', a reflection of its reputation as the only non-human enemy of the great whales.) However, it is the narwhal's most obvious feature – implicit in its binomial, *Monodon monoceros* – that is its own emblem of melancholy.

The narwhal's tusk is actually an overgrown, living tooth which erupts to pierce its owner's lip on the left-hand side and spirals up to nine feet long, sometimes even longer, but for centuries it was identified as the horn of a unicorn, invested with magical powers. A medieval conspiracy grew up between its Arctic hunters and the apothecaries who passed this natural

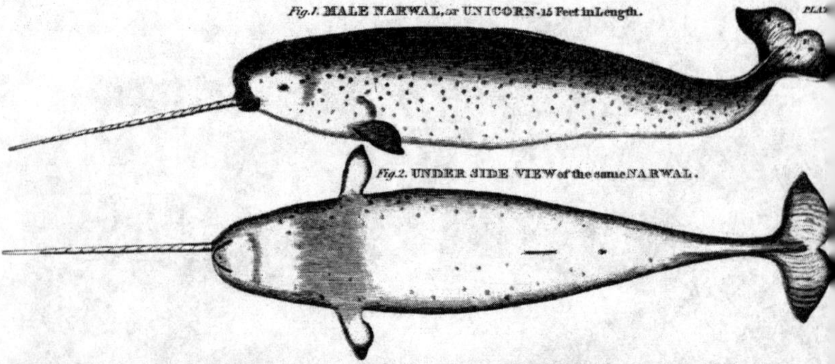

Fig. 1. MALE NARWAL, or UNICORN. 15 Feet in Length.

Fig. 2. UNDER SIDE VIEW of the same NARWAL.

wonder off as something truly legendary. Worth twenty times their weight in gold, the tusks were prized booty in the Middle Ages, stolen by Crusaders and traded across Europe as talismans of state. Only fifty were known to exist during the mid-1500s, and on his return from an expedition in 1577 to find the fabled North-West Passage, Sir Martin Frobisher presented Elizabeth I with a 'sea-unicorne's horn' valued at £10,000, more than the cost of a new castle. Evidently the Virgin Queen saw royal potency in the tribute, for she used it as a sceptre.

The powdered tusk was prized, too, as an antidote for poison and for melancholy, 'the English malady' whose anatomy, documented by the reclusive clergyman, Robert Burton, Melville had studied. Likewise, Albrecht Dürer's enigmatic *Melencolia* of 1514, with its brooding angel and a comet soaring over a distant sea provides *Moby-Dick* with its secret code according to one modern writer, Viola Sachs: a hidden structure based on the magic square of four. Through this cipher, says Sachs, Melville unites his melancholy theme with the story of the biblical outcast Ishmael and in so doing, 'expresses his vision of the terrestrial origin of creation'.

Freighted with such symbols and conspiracies, the narwhal became a fantastical beast, redolent with its own melancholy, as if bowed down by its onerous extension. To hold a narwhal tusk requires two hands, and the great ivory spike feels like carved stone in the grip, an ornate weight out of a cathedral mason's yard, belonging to the same place as a gargoyle. Little wonder that fairy-tale unicorns were, and still are, portrayed with a narwhal's horn.

It was only in 1685, when Francis Willughby described the narwhal in his *Icthyographia*, that the fraud was exposed. Further evidence came in the shape of the animals themselves, confronting us with their reality. In the 1880s a narwhal swam up the Humber and the Ouse to York, a medieval apparition in the shadow of the minster; a few years later, a beluga was shot in the same waterway, trying to make its way back to the north. In 1949 a pair of female narwhals appeared as far south as Rainham, Essex, and the River Medway in Kent.

Nowadays, microscopic examination has revealed the real magic of the narwhal's tusk. Unlike other teeth, its surface has open tubules connected to inner nerves; it is, in effect, a giant sense organ, lined with ten million nerve endings to enable the animal to detect subtle changes in temperature and pressure. This may explain why narwhals raise their tusks above water, as if to sniff the air. Other research indicates that the tusk is not only a sensory probe, but may also be a transmitter or receiver of sound, and even of electricity. Such discoveries exceed the narwhal's mythical powers. Its legendary spike is no dead bone, but an enervated growth producing 'tactile sensations' which 'might be interpreted as pleasurable'. Males rubbing their tusks together were formerly thought to be duelling over females; clearly, this behaviour has other aspects. So sensitive are these appendages that if broken, the animal suffers such severe pain that, in a remarkably philanthropic gesture, another narwhal will insert the tip of its own tusk in the exposed space, and break off the end to plug the aching gap.

Given such facts, who could resist a narwhal, with its shadowed damask, shrouded in black and white and grey and brown, monochrome daubs on a painter's palette? Perversely, it is the animal's other end that I find most beautiful: its wonderfully ornate flukes, flowing from a central notch in an exuberant sweep to the tips and back in an ogee curve to the tail stock. They may look back to front, but they are made for performance as much as any spoiler on a sports car.

The reader may guess that I am inordinately fond of the holarctic whales. Like belugas, narwhals also change colour as they age. It is an improbable sequence. They are born light grey, a nursery colour that endears them to their mothers; as they approach maturity, they darken to a purplish black. This then separates into black or dark brown spots, so that young adults resemble leopards or thrushes; in old age these marks recede, revealing the white below, just as the fine hair of an elderly woman turns silvery grey, making them seem wise as well as old.

This transformation is often thwarted by fishing nets or Inuit harpoons. The narwhal's blubber is a particular delicacy, and when a harpooned animal is hauled out of the sea, slices are eaten warm from its carcase as *mak taq* – a vitamin-rich fast-food to forestall scurvy. The Inuit carve its tusks into decorative objects, a useless embellishment for a thing of natural beauty. Yet to them, the narwhal is an entirely utilitarian catch: its lance makes fishing rods and its intestines supply the line; its fine oil is burned in moss lamps. In the past, both the narwhal and beluga have furnished soft leather for gloves, pale grey, white or mottled, ready decorated for a dandy's hands. One Hull company manufactured beluga bootlaces, with a somewhat self-defeating warning on the box, 'should not be pulled or jerked violently'.

In the mid-twentieth century, Canada imposed licences for hunting belugas, although native peoples and the Royal Canadian Mounted Police were still allowed to kill them 'for their own domestic use and for feeding their dogs'. Thousands of narwhals and belugas are still hunted every year from small boats or shot at from the ice, a cull in which nature itself is complicit. In winter, the inlets up which the animals swim can freeze over, creating a barrier too wide for them to cross in one breath. The whales are sealed in a blue-green world, one that threatens to become their collective tomb.

It is a heart-rending notion. At Point Barrow, Alaska, nine hundred belugas were forced to share an ice hole or *savssat* one hundred and fifty yards long by fifty yards wide. Unable to find open water, the animals surfaced every twelve to eighteen minutes, taking ten or fifteen breaths, then dove again, singing their distress. Their innate sense of community compounded the crisis as they rose to respire at one and the same time, a deadly synchronicity that caused some animals to be bodily squeezed out of this frozen hole of Calcutta and into the arms of the Inuit. In one day, they took three hundred whales.

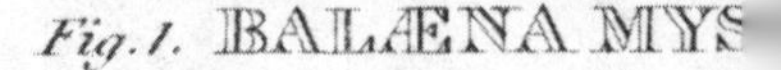

Fig. 2. CUB of the COMMON WHALE 17 Feet lo

But of all the Arctic cetaceans, the bowhead or *Balæna mysticetus* is the most mysterious. It is perhaps my favourite whale, although I have never seen one, and probably never will. Closely related to the right whale – distinguished mainly by its lack of callosities – it is able to break through the ice with its massive bow-shaped head, thereby avoiding the pathetic fate of its lesser cousins. It also has the longest baleen of any whale, measuring up to fifteen feet in length. Hanging in crystalline water with its huge white jaws decorated with a 'necklace' of black spots, this ebony-grey giant seems to embody the silent, ominous spirit of the Arctic – although, like the humpback, it too sings a low,

·COMMON WHALE.

on of the Whalebone.

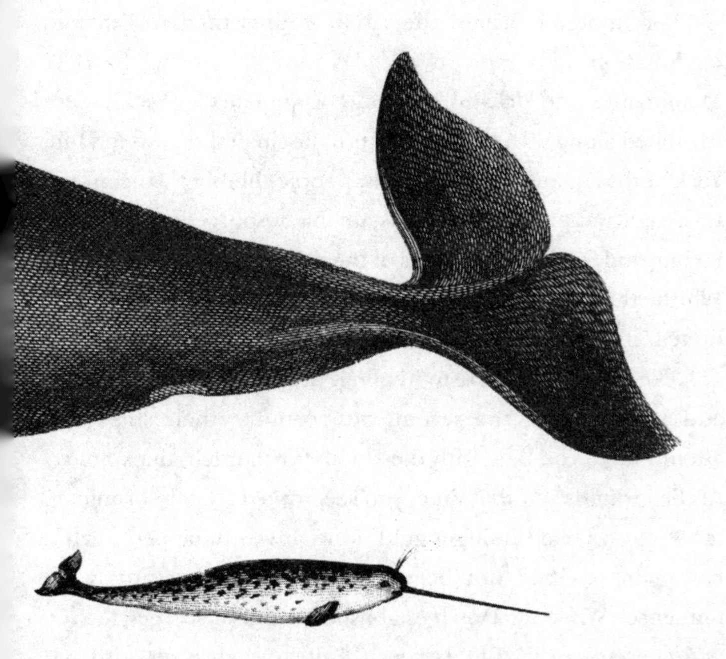

NARWAL. Length exclusive of the Tusk 14 Feet.

resonant song. Living at the top of the world, it is the first whale, one that struck even hardened whalers with a kind of awe. In 1823, the crew of the *Cumbrian* from Hull watched in fearful wonder as a fifty-seven-foot female bowhead circled their ship, then calmly pushed the vessel backwards with her snout to repel their invasion. For centuries the bowhead lived in icy obscurity; that was its salvation. Preserved by the very harshness of its environment, this vast creature simply vanishes when winter closes over the pole as though disappearing from a radar screen, upending its lacquer flukes and slipping back into the sea, along with its secrets. It has good reason to seek such sanctuary: the blubbery, baleen-heavy creature has learned to its cost that there is no hiding place so remote that it cannot be sought out by man.

For imperial Britain the Arctic represented wealth and exploitation, and even its peoples were fair game. In 1847 Memidadluk and Uckaluk, 'The Two Esquimaux, or Yacks', were exhibited along with their artefacts to fascinated crowds in Hull, York and Manchester. Fish, flesh, people, blubber, baleen, oil: the Arctic was an index of unsustainable resources ready for the taking, and for the inhabitants of the northern ports of Hull and Whitby there seemed to be an invisible tie between their maritime fastness and the frozen seas beyond.

The British came late to whaling. At the end of the sixteenth and beginning of the seventeenth century, their ships had attempted to compete with the Dutch for the rich, unexploited Arctic grounds: 'At that time, you see,' noted a later chronicler, 'whaling was like finding a gold mine. It was untapped wealth; the mammals had not been scared, and the rewards were immense.' While the Dutch established their Spitsbergen factory at Smeerenberg, or Blubbertown, British whalers set out from Hull and even from Exeter. But their trade declined in inverse proportion to Dutch success; by 1671, the Netherlands was sending out 155 whalers to Greenland, and sometimes their annual catch would reach as many as two thousand whales. In 1693 there was a move to revive the British industry 'formerly . . . very beneficial to this Kingdom', as Sir William Scawen, London financier and merchant, told Parliament, 'not only for the great Quantities of Whalebone and Oil which hath imported from thence; but also a Nursery for Seamen, and the Expence of Provisions for victualing the Ships'. Scawen bemoaned that since 1683, 'there hath not been one Ship sent from *England* to *Greenland*; so that Whalebone, which . . . was sold at Sixty Pounds *per* Ton, is now sold for Four Hundred Pounds the Ton; whereby *Holland* and *Hamburgh* draw out of this Kingdom above One hundred thousand Pounds for Whalebone and Whale Oil.'

Soon enough, business turned back to the whale. In the 1720s the South Sea Company, recovering from the infamous financial scandal of the Bubble, invested in whaling on the advice of Henry Elking, who too had bemoaned Britain's lack of initiative to be 'a very great Mistake'. The company fitted up a dozen ships on the Thames and sent its fleet north, encouraged by a government tax exemption for all whale products. The rewards were discouraging – the squadron returned with just twenty-five whales, barely enough to cover the cost of the expedition – and it was not until the mid-eighteenth century that the country applied itself seriously to whaling. Once begun, however, Britain excelled, employing the same efficiency it brought to slavery (in which my mother's own ancestors, living in Bristol, were complicit). On both were built the foundations of empire: the trade in humans, for sugar; and in whales, for oil.

As a result, London became the best-lit city in the world. By the 1740s, five thousand street lamps were burning whale oil, expunging the primal darkness. The capital itself was a whaling port. Unlike the Yankee syndicates, entire English fleets were owned by one merchant such as Samuel Enderby, Thomas Sturge, or Elhanan Bicknell. Their ships sailed from the Howland Great Wet Dock at Deptford, the largest commercial dock in the world, a huge gash cut into the south side of the river, precursor of the time when London would be undermined by its own commerce, its river banks riddled with such inlets. Able to handle one hundred and twenty ships, it was renamed Greenland Dock in honour of its Arctic trade, with quayside bollards made of whale bones. Try-works were established here too, where whales were processed away from the city so as not to offend its inhabitants with the stench. Further rendering was done around the looping bend of the river, on what became the site of the Millennium Dome. Here, where the coffee-coloured Thames widens into the sea, dead whales were brought back to London's streets. Here, where expensive Docklands flats now preside, blubber was also boiled.

It was on the eastern coast that British whaling truly prospered, however, from ports closer to the northern whale fisheries; pre-eminently Hull and Whitby. They had long experience of the earliest traditions of whaling: a thousand years before, the Vikings had whaled off Norway – in the saga of Beowulf, the sea is called a 'whale-road' – and by the ninth century were exporting whale meat to England. Eight hundred years later, in 1753, whaling began in Whitby. Only three animals were taken that season, but over the next eighty years, fifty-eight ships sailed on 577 voyages from the Yorkshire port, harvesting a total of 2,761 whales, 25,000 seals and 55 polar bears.

It was hardly a safe occupation for the hunters. During its peak years of whaling, Whitby lost seventeen ships – an awful attrition, added to by such individual tragedies as the death of four men when a boat of the unhappily named *Aimwell* was stove in by a whale in 1810. None the less, whaling was now a lucrative British trade, and by 1788 *The Times* was reporting munificent catches for the northern ports. In one week alone, the *Albion* brought into Hull '500 butts of oil and two tons of fins, the produce of seven and a half whales'; the *Samuel* arrived in the same port with '60 butts of blubber and one ton of fins, the produce of three whales'; and the *Spencer* arrived in Newcastle with '270 butts of blubber, and five and a half tons of fins, the produce of seven whales' – not including a further four ships bringing the bounty of sixteen and a half whales, and two thousand seals. The 'great slaughter of the Greenland whale' was truly under way, as techniques were improved to satisfy Britannia's need for oil to light her subjects' way, and for baleen to corset her Prince Regent in a 'Bastille of Whalebone'.

Like the Yankee whalers, the British hunted their prey from smaller boats, modelled, in their case, on early Viking craft. Whales were often killed from the ice, too, and dragged onto it to be butchered; unlike the Americans, British whalers did not render down blubber on board, but brought it back wholesale. So many ships were engaged in the business that up to a hundred vessels could be seen along the ice margin, a virtual cordon making it almost impossible for any whale to escape. It was nearly as hazardous for the whalers – one in ten of the ships would never return.

As war with America forced Britain to find new supplies of sperm oil, the government offered bounties of up to £500 to ships owned by companies such as Enderby and Sons. Samuel Enderby had arrived in London from Boston, Massachusetts, in 1775. He was a British loyalist – his ships had carried the famous consignment of tea into Boston Harbour. In 1776, along with Alexander Champion and John St Barbe, Enderby equipped twelve whalers with American captains and harpooneers. They returned with 439 tons of oil.

In 1788, acting on information gathered by James Cook, who had seen sperm whales on his voyage to Australia, Enderby sent out *Amelia*, the first British ship custom-built for 'sperming' to sail into the Pacific, thereby stealing a march on the Yankees, whose first ship, the *Beaver*, did not leave Nantucket for the Pacific until 1791. With on-board try-works enabling vessels to hunt far from home, these were 'by far the longest of all voyages now or ever made by man', as Ishmael says. They were the starships of their day, boldly going after animals whose own ancestors had colonized remote seas millions of years before. Now humans were creating their own new routes of oceanic colonization.

It was a new rivalry of the high seas. Through whaling, the British Empire extended its influence into the southern hemisphere in an 'atonement' for the loss of its American colonies. Britain intended to be self-sufficient in the matter of whale oil. 'We are all surprised, Mr Pitt,' a sardonic John Adams, the first ambassador of the new republic, told the Prime Minister in 1785, 'that you prefer darkness and consequent robberies, burglaries and murders in the streets to the receiving, as a remittance, our sperm oil.' Adams, a future president, spoke with the confidence of a former charge who had stolen a march on his master, 'seeing that the Yankees in one day, collectively, kill more whales than all the English, collectively, in ten years', as Ishmael boasts. Whaling was a presentiment of a new world order.

Whale-ships cleared the way for missions to the South Seas; whalers brought God as well as light to the world. As Hal Whitehead remarks, 'They left behind diseases, non-native animals (especially rats), technology, and their genes.' Outgoing British whale-ships – which would otherwise be empty – supplied the convict settlements of Australia. '*Evidence* inclines us to believe that these colonies would never have existed had it not been for whaling vessels approaching their shores,' Thomas Beale wrote. 'It is a fact, that the original settlers at Botany Bay were more than once saved from *starvation* by the timely arrival of some whaling vessels.' In 1791 the enterprising Samuel Enderby opened an office in Port Jackson, Sydney Harbour, and arranged for his ships to carry convicts there, delivering new slaves to New South Wales. The establishment of these colonies gave Britain a great advantage in the southern whale fisheries; soon those same colonies would be supplying sperm oil in their own right, hunting the animals from their own shores and exporting the products 'at a much less cost of time and capital' to Britain. Meanwhile, James

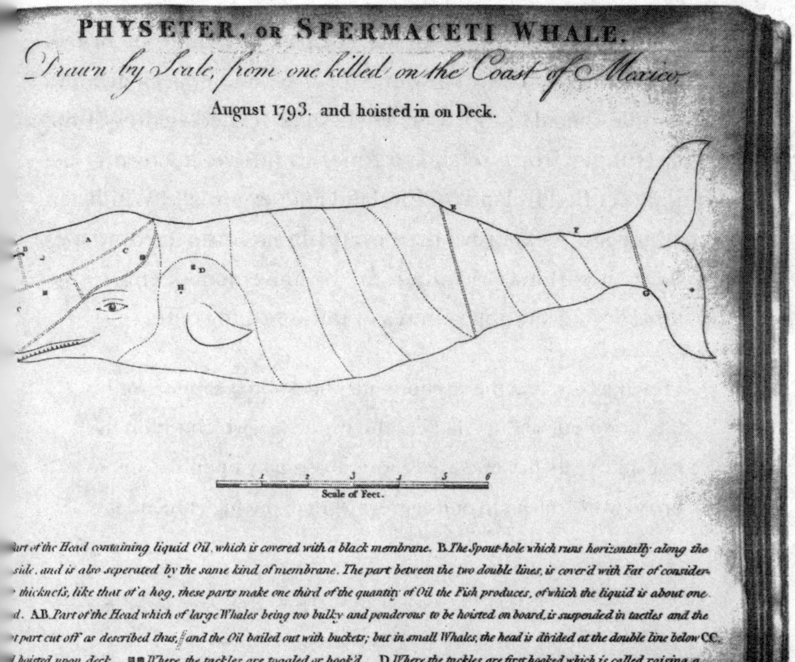

PHYSETER, OR SPERMACETI WHALE.

Drawn by Scale, from one killed on the Coast of Mexico,

August 1793. and hoisted in on Deck.

Scale of Feet.

art of the Head containing liquid Oil, which is covered with a black membrane. B.*The Spout-hole which runs horizontally along the side, and is also seperated by the same kind of membrane. The part between the two double lines, is cover'd with Fat of consider- thickness, like that of a hog, these parts make one third of the quantity of Oil the Fish produces, of which the liquid is about one d.* AB.*Part of the Head which of large Whales being too bulky and ponderous to be hoisted on board, is suspended in tactles and the t part cut off as described thus, and the Oil bailed out with buckets; but in small Whales, the head is divided at the double line below* CC. *hoisted upon deck.* *Where the tackles are toggled or hook'd.* D *Where the tackles are first hooked, which is called raising, a ce, being thus steadied in the tackles the head is divided at the lowest double line and wore a stern till the fish is flinched, which is e by seperating the Fat from the Body with long handled Iron Spades, as the Whale is hove round by the tackles the Fat peels and if any Sea is on the rising of the Ship considerably expedites the business.* — E.*A large lump of Fat.* F.*A smaller.* — *n the Fish is flinched, or peeled to* E. *it will no longer cant in the tackles, is therefore cut through at the first double line and also* G. *the Tail being of no value.* — H.*The Ear, which is remarkably small in proportion to the body, as is also the Eye from which a llow or concave line runs to the forepart of the head the Eyes being prominent enables them to pursue their Prey in a direct line, d by inclining the head a little either to the right or left to see their enemy a stern, they have only one row of Teeth, which are the lower Jaw with sockets in the upper one to receive them, the number depends on the age of the Fish, the lower Jaw is a solid e that narrows nearly to a point and closes under the upper, when they spout, they throw the water forwards and not upwards like er Whales except when they are enraged, they also spout more regular and stay longer under water the larger the Fish the more quently they spout and continue longer under water. The Tail is horizontal with which he does much mischief in defending him- f. Their Food, from all the observations I have had an oppertunity of making, has been the Sepia or middle Cuttle Fish.* ——— *s species of the Whale, is remarkable for its attachment and for assisting each other when struck with a harpoon: and more mis- f is done by the loose Fish, than those the boats are fast to, and they frequently bite the lines in two which the struck Fish is fast with* — *e Ambergrease is generally discover'd by probing the intestines with a long Pole, when the Fish is cut in two at* E. ———

LONDON: Published January 1st 1798, by A. ARROWSMITH, Charles Street, Soho.

Colnett, an officer of the Royal Navy, sailed from Portsmouth on HMS *Rattler* to extend the nation's whale fisheries in the Pacific, although Ishmael mocks his rendition of a whale: 'Ah, my gallant captain, why did ye not give us Jonah looking out of that eye!'

More than ever, whaling was seen as 'the mine of British strength and glory', a vital source of maritime experience and mercantile speculation. Later, whale-ships ferried victims of the Great Hunger from Ireland to America, just as my own great-grandfather fled Ireland for England and, eventually, Whitby. In a manner more extensive than even Ishmael suspected, whales played their part in world affairs, in the movement of entire populations, and in shifting spheres of influence to come.

> I freely assert, that the cosmopolite philosopher cannot, for his life, point out one single peaceful influence, which within the last sixty years has operated more potentially upon the whole broad world, taken in one aggregate, than the high and mighty business of whaling.
>
> The Advocate, *Moby-Dick*

Each April, when the weather improved, Whitby's ships set off for Greenland, harpooning the easily caught whales of the Arctic. They brought back chunks of blubber, creating a stench that Ishmael compares to a whale cemetery, and which turned the port into one of the most noisome places in England.

Whitby whaling captains – many also Quakers – built their elegant houses high on the West Cliff, out of range of the stinking manufactory of their fortunes. Their Georgian terraces still overlook the harbour, a view framed by Whitby's famous whale bone arch. When I stood under this monument as a boy, I presumed it had been there for centuries. In fact, its mandibles, from a blue whale, were only erected in 1962, and have since been replaced by the jaws of a bowhead, presented by the people of Alaska. But down in the town, whale bones were used as timbers for roofs and walls. Entire houses and workshops were constructed from these giant ribs and jaws. If a man might stand within a whale's mouth, why not make a more convenient shelter for himself and his family, swapping bones for bricks? After all, the whale had no need for them now.

The expansion of the whaling fleet was almost frightening in its speed. In 1782 there were forty-four British whale-ships operating in the Greenland Sea. Two years later, that figure had doubled; and by 1787, 250 ships were sailing from British ports – only for a new war to erode their returns; 'greenmen' were imposed on whale-ships as a means of training them for the navy, while experienced whalemen were in turn impressed to fight Napoleon.

As a young sailor, William Scoresby had himself been captured off Trafalgar, and in a daring escape, managed to evade his Spanish jailers, stowing away on a British ship exchanging prisoners of war. Back in Whitby, Scoresby enlisted on the whale-ship the *Henrietta,* rising quickly to the rank of Specksioneer (a Dutch-derived term for principal harpooneer), then to captain. It was the beginning of a career that would claim the lives of no fewer than 533 whales.

Scoresby was a powerfully built man of great vitality, and his talent for whaling was undeniable. On his second voyage as master of the *Henrietta,* he returned with eighteen Greenland whales; in the next five years, the ship took eighty more animals, yielding nearly 800 tons of oil. Soon Scoresby was commanding a larger ship, the *Dundee*; on her first voyage she garnered an unprecedented thirty-six whales. Scoresby's heroic status was only reinforced when his ship faced a French warship off the Yorkshire coast and destruction seemed imminent: at the last moment the *Dundee* uncovered her eighteen-pound guns, at the sight of which the enemy turned and fled.

In 1803 Scoresby took command of a new, double-hulled vessel with metal plates on her bow, enabling her to plough through Arctic ice. With the *Resolution* sailed his own son, also named William, fourteen years old and about to become a whaler, explorer and inventor in his own right. He had a heady precedent to follow. Scoresby Senior had come closer than any other man to claiming the £1,000 prize offered to anyone able to sail north of the eighty-ninth parallel in pursuit of the fabled North-West Passage, a search which 'laid open the haunts of the whale'. He had also devised an enclosed crow's nest, an ingenious contraption with a protective framework of leather or canvas, storage space for telescope and firearm, and flags and speaking trumpet for communication with the crew or other ships. It was an eccentric device to Ishmael, who satirizes Scoresby as 'Captain Sleet', standing in his invention, armed with a rifle 'for the purpose of popping off the stray narwhales, or vagrant sea unicorns infesting those waters'.

Scoresby was no ordinary seaman, often writing his ship's log in verse: 'So now the Western ice we leave/ And pleasant Gales we doe receive'. He also kept a pet polar bear, which he would walk on a leash down to Whitby harbour to fish for its lunch. Scoresby presided over the peak years of British whaling, the personification of this national harvest. By the summer of 1817, columns in *The Times* were being devoted to reports from

Berwick, Greenock, Peterhead, Aberdeen, Montrose, Dundee, Kirkcaldy, Leith, Liverpool, Hull, Newcastle, London and Whitby as whale-ships returned laden with blubber and bone.

In 1823, after a long and successful career, Scoresby gave up the sea and retired to Whitby. He had never questioned the right of man to take the whale; rather, he reasoned that whaling was a tribute to man's ingenuity and God's grace. 'We are led to reflect on the economy manifestation respect to the hugest of the animal creation, whether on earth or in the ocean, whereby all become subject to man, either for living energy or the produce of their dead carcasses.'

'The capture of the whale by man, when their relative proportions are considered, is a result truly wonderful,' Scoresby declared. 'An animal of a thousand times the bulk of man is constrained to yield its life to his attacks and its carcase a tribute to his marvellous enterprise.' His was a righteous pursuit, 'satisfactorily explained on the simple principle of the Divine enactment. It was the appointment of the Creator that it should be so.' However, Scoresby's own departure from the world would be as brutal as his butchery of the whales of the Arctic.

An elevated pavement runs the length of Bagdale, passing the elegant terrace that overlooks a walled Quaker graveyard and, beyond that, Pannet Park, once decorated with whale bone arches, as were many other Whitby gardens. Here Scoresby lived, in a fine Georgian house with classical fanlights and carved sandstone porch. And here, on 28 April 1829, at the age of sixty-nine, he took up his pistol, and shot himself through the heart. 'He appears to have been in a state of temporary derangement for several months past,' the subsequent inquest found. It is impossible to know the reason why Scoresby took a pistol to his heart; it would certainly be too sentimental to read into his self-murder any sense of guilt over the five hundred whales he had killed, and for whose deaths he gave thanks to God.

From his portrait, William Scoresby appears a refined version of his father; rational, scientific, pious, inquisitive; a combination of disciplines that the era allowed. He too had gone to sea as a boy, but had studied science in Edinburgh before joining the navy, and after his discharge went to London to meet Joseph Banks, the renowned naturalist who had sailed with Captain Cook. Banks received Scoresby at his house in Soho Square; he may have seen something of himself in this young man who spoke so eloquently of the Arctic regions he had already explored on his father's whaling voyages.

A year later, William took command of his first whale-ship, the *Resolution*, followed by the *Esk*. On his voyages, he sought to prove that the temperature of the sea was warmer below its surface. Sending his findings to Banks, the two men developed an instrument to more accurately measure the ocean's residual heat: the 'Marine Diver', a brass contraption that could be lowered 7,000 feet into the water. For Scoresby, whaling was a way to finance his investigations. During the *Esk*'s perilous journey – when he nearly lost his ship to pack ice, and when his men may have cursed their captain's curiosity – Scoresby made scientific notes in books and

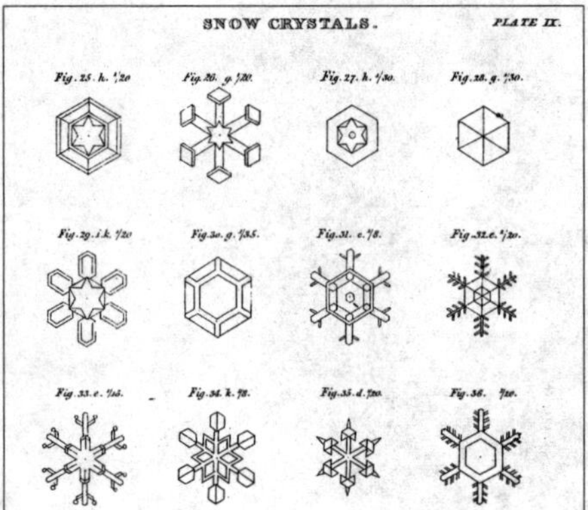

SNOW CRYSTALS. PLATE IX.

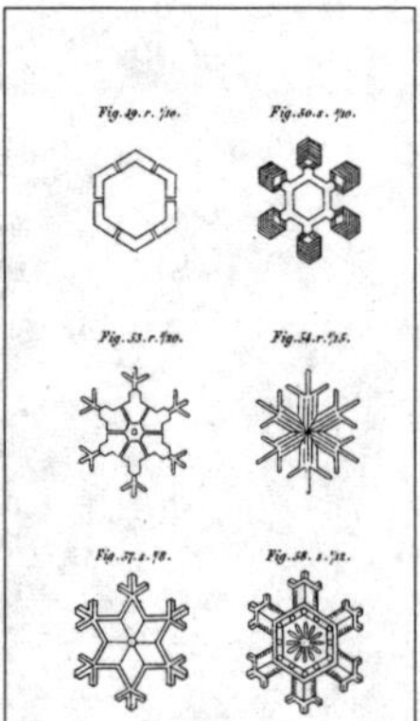

papers which flowed out over his desk: calculations, sketches, suppositions and descriptions, a fluid body of work which, for the first time, documented these unblemished seas.

An Account of the Arctic Regions with a History and Description of the Northern Whale-Fishery was published in two volumes in Edinburgh in 1820, and profusely illustrated with maps and engravings. Scoresby's became the work by which all others were measured, a compendium of cetology and whaling techniques and the nature of the Arctic itself, complete with ninety-six snow crystals illustrated by this son of Captain Sleet in dizzying pages of repeated patterns.

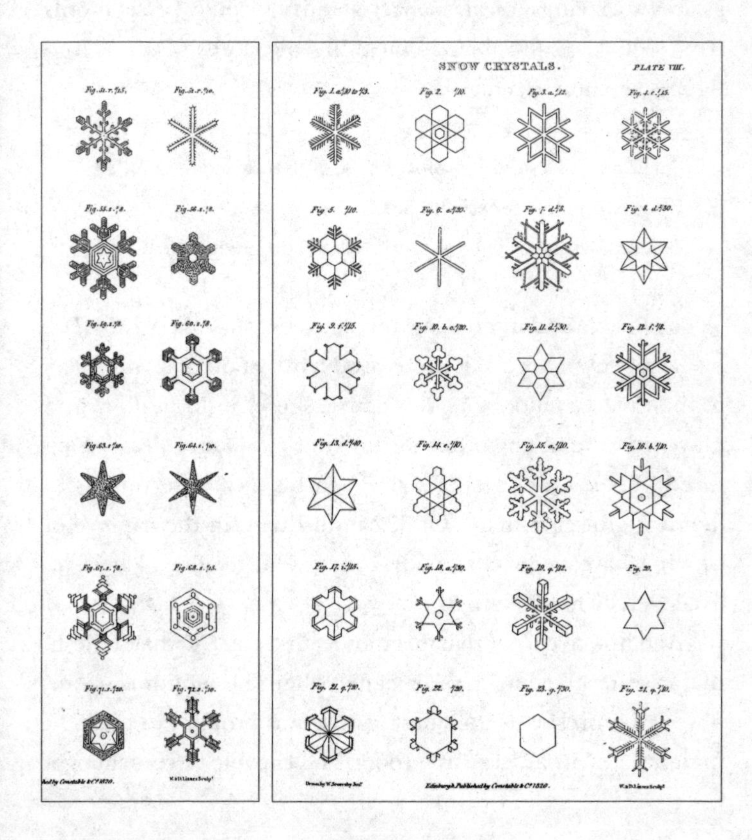

A polar counterpart to Beale's Pacific travels, Scoresby's text had a religious overtone, as if the animals and places and phenomena he catalogued were evidence of Eden; the book was later taken up by the Religious Tract Society and supplied to the American Sunday School movement as an affirmation of the Creation. For all its scientific rigour, there was no conflict between its author's beliefs and his investigations. Scoresby's faith echoed his father's; and like his father, his avowed aim, by God's grace, was to find the North-West Passage. But if the whale's true nature could only be guessed at from its spouting, steamy surfacings, or an iceberg's entirety could be seen only from below, so the deeper meaning of Scoresby's facts and figures lay submerged.

> Section I. A Description of Animals, of the Cetaceous Kind, frequenting the Greenland Sea.
> *Balæna Mysticetus*: The Common Whale, or Greenland Whale.

'This valuable and interesting animal, generally called *The Whale* by way of eminence . . . is more productive of oil than any other of the Cetacea, and, being less active, slower in its motion, and more timid than any other of the kind . . . is more easily captured.' Like Beale, Scoresby drew on his own observations to delineate the leviathan. 'Of 322 individuals, in the capture of which I have been personally concerned, no one, I believe, exceeded 60 feet in length . . .'

And how to convey that magnitude, that mass of whalish flesh, that cavern of ceiling-high baleen? 'When the mouth is open,' observes Scoresby, as the latest captive was brought to book, 'it presents a cavity as large as a room, and capable of containing a merchants-ship's jolly-boat, full of men, being 6 or 8 feet wide, 10 or 12 feet high (in front), and 15 or 16 feet long.' Such detail is as telling of its author and his time as it was of the whale. 'The eyes . . . are remarkably small in proportion to the bulk of the animal's body, being little larger than those of an ox,' he continues, writing at his desk by whale-light, 'nor can any orifice for the admission of sound be described until the skin is removed.' As with so much of the whale, so little could be discovered until it was dead.

Yet that tiny eye is all-seeing. 'Whales are observed to discover one another, in clear water, when under the surface, at an amazing distance. When at the surface, however, they do not see far.' In fact, they sensed one another's presence by sound, even though, like Beale's sperm whales, Scoresby considered bowheads to be dumb. 'They have no voice,' he concludes, 'but in breathing or *blowing*, they make a very loud noise.' The ice echoes to these trumpets of watery elephants, effortlessly negotiating oceans that defeated mere unblubbered men. 'Bulky as the whale is, as inactive, or indeed clumsy as it appears to be . . . the fact, however, is the reverse.'

And age, Scoresby, sir, what of that? 'In some whales, a curious hollow on one side, and ridge on the other, occurs in many of the central blades of whalebone, at regular intervals of 6 or 7 inches,' replies the captain testily, somewhat irritated at my interruption. 'May not this irregularity, like the rings in the horns of the ox, which they resemble, afford an intimation of the age of the whale?' It took science two hundred years to catch up with another of Scoresby's discoveries, one that lay buried, almost unnoticed, in his book. In his search for the fabled North-West Passage, he had, accidentally, opened the way to the whale's most abiding mystery.

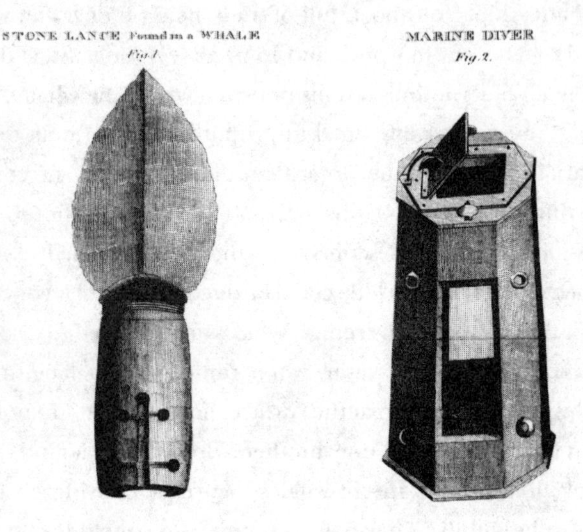

Set with deliberate anachronism next to Scoresby's technological Marine Diver is a drawing of a stone tool, a Neolithic contrast to an invention of the Industrial Revolution. 'The master of the Volunteer, whaler of Whitby, when near the coast of Spitsbergen, July 19. 1813, shewed me part of a lance which had been taken out of the fat of a whale killed by his crew a few weeks before,' Scoresby related, with a degree of cool amazement. 'It was completely embedded in the blubber, and the wound was quite healed. A small white scar on the skin of the whale, alone marked the place where the lance had entered.' But the telling fact was that such weapons were 'in common use among the Esquimaux a century ago'.

Scoresby found that these tools were 'struck by some tribe of the same nation, inhabiting the shores of the frozen ocean, on the northern face of the American Continent, yet unexplored'. If whales had been caught in the Atlantic implanted with implements made on the Pacific coast like some early form of tracking device, then there must be a passage between the two oceans. (Three centuries earlier, when Sir Humphrey Gilbert presented his argument for a North-West Passage to Elizabeth I – a year before Frobisher's expedition – he cited as evidence a narwhal horn found on the Tartary coast.) This was the Holy Grail for which Scoresby and his father had searched, the opening up of the northernmost world. In the pursuit, they were diverted from a yet more extraordinary finding: Scoresby's Marine Diver may have plumbed the waters to reveal their depths, but this primitive artefact had uncovered the sensational secret of the bowhead.

On 29 April, 1850, Herman Melville withdrew the two volumes of Scoresby's work from the New York Society Library. As he read the *Arctic Regions* – which he failed to return for a year – Melville's imagination was fired by the story of the stone lance. It led him to a startling conclusion. In *Moby-Dick*, Ishmael reports the finding, during the cutting-in of a whale, 'of a lance-head of stone . . . the flesh perfectly firm about it. Who had darted that stone lance? And when? It might have been darted by some Nor'West Indian long before America was discovered.' Even given his exaggeration, the idea was staggering: if Scoresby's lance was a century old, then this meant the animal was even older.

Until recently, fact-checking editors have dismissed Ishmael's airy claims. 'Anatomical evidence from larger whales suggests a life of up to seventy or eighty years,' Harold Beaver assured readers in a footnote to *Moby-Dick* in 1972. 'But a longer span, stretching to centuries, is sailors' myth.' Now, in a belated confirmation of Melville's musings, scientists are beginning to realize that ideas about how long whales may live have been substantially underestimated. The clues to this revision came from native Alaskans who still hunt bowheads in the Bering Sea. The Iñupiat have observed the whales for centuries, and their storytellers claim to have recognized individual animals over successive human generations. Since 1981, six stone or ivory harpoon points have been found in the blubber of whales – weapons that modern Iñupiats did not recognize, having used mostly metal harpoons since the 1870s.

Long after Scoresby's discovery, scientists came to their own conclusion: that these whales must be as old as the implements found in them. And as the Iñupiat hunted only young whales, being better eating, it seemed likely that there might be even older animals, hidden in the icy reaches. The bowhead's Arctic existence seemed to slow down its life, extending it, decade by decade, century by century; a sentient entity suspended in vast vistas of time, virtually cryogenically preserved.

Using a technique for dating animals from changes in the aspartic acid levels in their eyes, Dr Jeffrey L. Bada of the Scripps Institution of Oceanography in California examined tissue from whales caught by the Iñupiat hunters. Most were twenty to sixty years old when they died; but of five large male bowheads, one was ninety years old, four were aged between 135 and 180 years, and one was 211 years old. Employing other methods for measuring radioactive lead in bone, and samples of skin collected from living whales, Dr Bada stated that 'what we have assigned the bowheads are only minimum ages . . . These are truly aged animals, perhaps the most long-lived mammals.'

Since it is unlikely that the oldest whales have been caught – that older whales could and most probably do exist – Bada's assessment is hardly underestimated. Even as I write, a three-and-a-half-inch lance tip, made in New Bedford in the 1890s, has been retrieved from the blubber of a bowhead caught off Alaska. The consequences haunt me: that these whales swam the same seas that Scoresby had negotiated; that the same animals from which he had made his observations might yet be alive. It is also an exquisite revenge: born before Melville, the whales have outlived their pursuers.

In his chapter entitled, 'Does The Whale's Magnitude Diminish? – Will He Perish?', Ishmael dismisses the idea that whale populations, particularly those of the great baleen species, were declining. On the contrary, he claimed, they had 'two firm fortresses', which, he averred, would 'for ever remain impregnable . . . their Polar citadels . . . in a charmed circle of everlasting December'.

> Wherefore . . . we account the whale immortal in his species, however perishable in individuality. He swam the seas before the continents broke water; he once swam over the site of the Tuileries, and Windsor Castle, and the Kremlin. In Noah's flood he despised Noah's ark; and if ever the world is to be flooded again, like the Netherlands, to kill off its rats, then the eternal whale will still survive, and rearing upon the topmost crest of the equatorial flood, spout his frothed defiance to the skies.

In his fantasy, Melville saw a new dispensation, a watery version of Hawthorne's prairie holocaust, Harry Hinton's icy sanctuary come to life.

It was recently announced that the earth entered a new geological epoch around the year 1800, as a result of the Industrial Revolution. The era of the Holocene has ended, say the scientists; the era of the Anthropocene has begun. In one of the cathedrals in the walled city where the tsars of all the Russias lie in their monumental sarcophagi, there is a medieval mural of a whale. This fragile image has seen off Peter, Nicholas, Josef and Mikhail, surviving human empires like some Neolithic cave painting, like the stone within the whale. Now the whale's citadels are rapidly receding, making the North-West Passage a permanent reality, opening continent to continent as fresh water locked into polar ice leaches into the ocean, and the world's northern nations prepare to plunder the Arctic's resources anew.

What will this mean for the whale, as the sea rises to remind us of its power? Krill, which feed on the algæ on the undersurface of the ice, may diminish, and food sources for the whales are already becoming scarce at lower latitudes as warming oceans push them ever further north, only to find that those everlasting citadels have vanished. On the other hand, the mineral nutrients released by the same process in the Antarctic may have beneficial effects for the food chain and, perhaps, cetaceans. No one really knows. We are living through a vast experiment, one which may result in the flooded world that Melville imagined; a world that the whales will inherit, evolving into superior beings with only distant memories of the time when they were persecuted by beings whose greed proved to be their downfall.

Having published his book, Scoresby returned to the sea in the newly built *Baffin*, taking leave of his wife and his family in Liverpool – had he but known it, for the last time. He returned home in September 1822, after charting the east coast of Greenland, to be told of his wife's death. Disheartened, he made only one more trip, before giving up the sea for another vocation: that of vicar. As its erstwhile champion exchanged 'the clatter of hailstones on icebergs' for the sound of psalms from pews, Whitby's whaling fleet dwindled to just ten ships. In 1825, at the parish church of St Mary's, high on the hill overlooking the town, it was Scoresby's sad duty to preach on the occasion of the loss of all hands on the *Lively* in an Arctic storm; and the *Esk*, his own former command, which sank just thirty miles from Whitby. The terrible toll of sixty dead added up to the end of an industry – as did the depletion of the whaling grounds, and the thousands of slaughtered animals.

Scoresby became Vicar of Bradford – where his parishioners included the Reverend Patrick Brontë of Haworth village, and his young daughters – and turned his scientific attention to the mysterious forces of mesmerism. Instead of dealing in oil and whalebone, Whitby now traded in jewellery carved from the shiny black jet found in its cliffs and made mournfully fashionable by a perennially grieving queen. And by the time my own grandfather was walking along Bagdale on his way to Mass with his brothers and sisters, Whitby's arched buildings of bone were dwarfed by the railway viaduct, overturned arks for a new age of extinction.

Fin whale beached at Winterton, Norfolk, in a storm, 5 January 1857. It was later exhibited in the Mile End Road, Whitechapel.

XI

The Melancholy Whale

> A tenth branch of the king's ordinary revenue . . . is the right to royal fish, which are whale and sturgeon. And these, when either thrown ashore or caught near the coast, are the property of the king.
>
> *Blackstone,* Extracts from a Sub-Sub-Librarian

On a bleak strand south of Skegness and its garish amusements, the sun was already beginning to set as I trudged through the damp grey sand. Something lay ahead in the creeping dusk, growing closer until its vague shape resolved into a discernible form. Before that, I smelled it. I can smell it still when I look at the pictures. Lying there, like a cod on a fishmonger's slab, was a minke whale. Its shiny black skin had been entirely flayed, leaving a fishy-coloured beige, the texture of latex – except where the blubber had begun to turn blue-green.

When I had last seen a minke, it was surfing over Stellwagen Bank, snatching breaths at the surface, briefly showing the sharp-pointed rostrum for which the whale is named, *Balænoptera acutorostrata.* (It owes its other name to a Norwegian sailor, Miencke, who mistook this, the smallest rorqual, for more valuable prey. Not for the first time, it occurs to me that whales are named for their usefulness to man, rather than for their innate beauty.)

Then, in a rare moment of revelation, a minke had swum by the bow of the boat, clearly silhouetted below, its fins emblazoned like the chevrons on an officer's sleeve. Now all I saw was a piece of dead matter that smelled like something between fish and meat. Its elegant flukes were reduced to raw cartilage; there was barely anything to indicate that it had ever been alive, save for its pale little penis hanging from the underside of its belly, flaccid and worm-like. I fingered it, then I walked back in the failing light, the moon rising like a bloody pearl out of the North Sea.

This stormy eastern coast has always been a wrecking place for the whales, forever echoing to their plaintive blows. Eighty years before my encounter in Skegness, another minke washed up near Mablethorpe, in September 1926. This time, the animal was still alive when it beached. Summoned from the Natural History Museum, Percy Stammwitz attempted to return the fifteen-foot female to the sea, without success, before claiming it as a specimen. Newspaper reports claimed that the whale lived for a day and a half after its capture, and that on its way to South Kensington, destined for the sand pits, 'its blowing was audible notwithstanding the noise of the engine, until the lorry was within about 30 miles of London, when the animal burst a blood-vessel and died of hæmorrhage of the lungs.' In truth, Stammwitz was aware that the whale was still alive when it was put on the lorry, but as it was not conscious, he reasoned that if he tried to kill the animal, it could suffer more if it regained consciousness in the process.

In 1913 the Crown gave first right of refusal to the carcases of royal fish to the Natural History Museum, thereby recognizing their scientific as well as their commercial value. The Board of Trade requested that the Receivers of Wreck – then stationed around the country – should send 'telegraphic Reports' of stranded whales to the museum. The earliest records, compiled during the years of the First World War, were collated by the museum's famous director, Sidney Harmer. They were sad casualty lists to mirror others being published at that time (when, as Harmer noted, coastguards had other matters to deal with): from a finback in the Firth of Forth, 'at first supposed to be an aeroplane', to a rare Sowerby's beaked whale found at Skegness and which 'appeared to have been killed by rifle-shots, perhaps in mistake for a German submarine'. The animal's calf lay alongside it on the beach. Other whales died after swimming into mine fields intended to blow up U-boats.

Thirteen thousand beached whales have been recorded by the museum, but as they span the twentieth century – each lost and expired cetacean plotted on a deathly map of the British coastline – only a few have been claimed for scientific research. The remainder represent a collective rebuke and a logistical dilemma, for even in death, whales present humans with gargantuan problems.

From a modern office block overlooking Southampton Docks, the Receiver of Wreck administers a fourteenth-century decree. Since 1324, when the right was enshrined in the reign of Edward II, every whale, dolphin, porpoise and sturgeon found on English shores has become the property of the monarch. What was once a royal prerogative is now a liability. In the twenty-first century the Receiver of Wreck is, in effect, whale undertaker to Her Majesty Queen Elizabeth II.

The Receiver, or her Deputy – the current holders of these ancient posts both happen to be young women – will be alerted by one of nineteen coastguard stations. A dead or dying whale might be floating at sea, a potential shipping hazard, or it may present a public nuisance as it is washed up. Sometimes a whale will appear on one beach, only to be carried by the tide to another. In this morbid game, it is the Receiver's job to deal with the equivocal prize: a massive, stinking carcase. In remote locations, the whale is allowed to become carrion for birds; elsewhere, police cordons may be needed – less to shield people from zoonotic or interspecies infection than to protect them from the heavy plant machinery required to move an animal weighing many tons.

These are expensive disposals. Small whales cost from £6,000 to £8,000 to shift; large whales as much as £20,000. A profitable right has become a public expense. When whales were unprotected, they were valuable commodities, bounties to be claimed by the Crown; now they are treated as managed or even toxic waste, a result of pollution, or of the large doses used to euthanize the animal. And although they soon decay – the epidermis peeling, the internal organs breaking down, swelling their bellies with gas – dead whales remain resilient. Their blubber is thick and hard to puncture, and carcases hang from mechanical claws like Indian mystics suspended from hooks. Sometimes a pair of excavators join forces to pull them apart; other techniques include the use of high-pressure water jets. One fin whale that stranded on the Isle of Wight, having drifted from the Bay of Biscay, required nine truckloads to cart it away piecemeal to the local landfill. Another, at Lee-on-Solent, was interred at a site in the New Forest.

In her open-plan office, Sophia Exelby shows me her gallery of stranded whales. It is a gruesome car-insurer's album, each cetacean crash more ghastly than the last. A pilot whale lodged in Devon rocks, caught in the kind of boulders over which children clamber looking for rock pools. A finback washed up at Ventnor, its blubber dripping like wax in the sun, its separated head yards down the shore. A sei whale, one of the rarer rorquals – and one of the fastest – lying on Morecambe Sands, victim of its deceptive tides. A humpback in Kent, slumped on its white flippers like an airliner in an emergency landing. An orca in the Mersey.

Whales where they should not be.

Many may be accidents, such as ship-strikes, or the result of entanglements or disease. Mass strandings – more familiar on the beaches of Cape Cod or New Zealand – are less easily explained. Freak tides, bad weather, sand bars and ailing whales inadvertently leading their fellows to disaster have all been cited as possible causes; in his notes on strandings, Sidney Harmer remarked that they often occurred when the sea temperature was abnormally high or low, due to an influx of water from colder or warmer latitudes, and that a localized wind was blowing onto the shore.

Another theory is that whales align themselves to the earth's invisible power lines by means of magnetosomes in their bodies; ferro-magnetic material has been found in the tissues of cetacean organs to support this idea. Ever aware of their position – birds may use a similar technique in their migrations – they orientate themselves to magnetic contours as though possessed of personal GPS systems. But sometimes there are anomalies in this unseen map, lines running at right-angles rather than parallel to the land, or places where the coast has changed and no one has updated their system.

For a marine mammal, such a mistake can be fatal. The Cape's sands – laid down since the ice age – are a case in point. Deceived by their own senses, pilot whales and dolphins are led onto shore rather than through deep water. Spurn Point – a kind of Cape Cod in miniature at the mouth of the Humber – may have the same effect. Even more recent research shows that increases in strandings may also coincide with solar activity which disrupts the magnetic field. Studies of sperm whales stranded in the North Sea over the past three centuries indicate that ninety per cent occurred when the sun's activity cycle was below average – a finding that raises the notion that those seventeenth-century Dutch omens of catastrophe might have a meteorological, as well as an eschatological basis.

Other reasons put forward for mass strandings raise intriguing questions about the whales themselves. One biologist believes that such behaviour is a genetic memory of their evolutionary past: that stressed and ailing whales seek to return to the land because they know that at least they will not drown there. Some see a Malthusian instinct for the preservation of the greater species: mass strandings as a kind of population control at times when the whale numbers in a certain area have reached their sustainable limit. The fact that strandings increased after the end of commercial whaling is given as evidence for this rather drastic self-restriction.

However, decidedly unnatural forces may act as siren voices. It is increasingly certain that whales are affected by powerful military sonar, developed since the 1960s to detect newly silent enemy submarines. Strandings have been noted near naval exercises, during which sounds twice as loud as a jet engine are created. Toothed whales, reliant on their own sonar, are particular victims of this distortion of their natural soundscape; worst affected are beaked whales, which must normally rise slowly after their deep dives. The loud pulses panic them into surfacing, and gas bubbles form in their bloodstream, inducing compression sickness. Necropsies also indicate massive hæmorrhaging around their brains and spinal cords.

Anthropogenic noise may be the reason for the frequency of modern strandings on the British east coast, where seismic soundings for oil surveys not only cause localized distress, but may disturb the whales in their ancient sonic routes, causing them to take a wrong turn into the unsuitably shallow North Sea where they fail to find adequate food. Or it may be (as ever with whales, there are so many maybes and precious few certainties), as Liz Evans-Jones, who oversees the strandings project at the Natural History Museum, told me, that modern incidents are more likely to be reported because people are aware of the animals' plight, and that once remote coasts are now accessible. Whatever the truth, encounters with man's world seldom turn out to be beneficial for the whale.

In the past, shore dwellers had regarded a beached whale as a gift from the gods; those less accustomed to such events saw a dead cetacean as an evil omen, like a comet or an eclipse. After a whale arrived in the Thames during a storm in 1658, it was taken to have been an augury of the demise of the Lord Protector, Oliver Cromwell, who died the following day. It was certainly a strange sight: a thrashing leviathan off Dagenham. In his diary, John Evelyn, whose estate overlooked the river, noted: 'A large *Whale* was taken betwixt my Land abutting on the *Thames* & *Greenwich,* which drew an infinite Concourse to see it, by water, horse, coach, and on foote, from *Lond.*, & all parts.'

Amazingly, this was a right whale, an animal more suited to waters rich in plankton rather than the floating detritus of seventeenth-century London. The whale first appeared at low water, 'for at high water, it would have destroyed all the boates'. The alien was doomed by its unlucky appearance, as if its ungainliness itself was a sin; cornered, it fought back in a manner with which whale rescuers would be familiar: 'after a long Conflict, it was killed with the harping yrons, & struck in the head, out of which spouted blood & water, by two tunnells like Smoake from a chimney; & after an horrid grone it ran quite on shore & died.'

An amateur scientist himself, Evelyn took the opportunity to measure the monster. 'The length was 58 foote: 16 in height, black skin'd like Coach-leather, very small eyes, greate taile, small finns & but 2: a piked snout, & a mouth so wide & divers men might have stood upright in it: No teeth at all, but sucked the slime onely as thro a greate made of that bone which we call Whale bone.' Evelyn found it wonderful 'that an Animal of so greate a bulk, should be nourished onely by slime'. Sixty years later, on his 1721 tour of Britain, Daniel Defoe recorded a whale bone arch on the London to Colchester road, 'a little on this side the Whalebone, a place on the road so called because the rib-bone of a large whale, taken in the river Thames, was fixed there in 1658, the year Oliver Cromwell died, for a monument of that monstrous creature'. Whalebone Lane still exists in Dagenham; the bones are preserved in a local museum.

Other would-be visitors to London were little better treated than Evelyn's whale. In 1788, twelve male sperm whales stranded and died along the Thames estuary, almost within sight of the Great Wen itself; they were soon boiled down for oil. Five years later, in an event recorded by Joseph Banks, a thirty-foot orca entered the river and found itself the subject of 'an exciting chase' after it was harpooned, towing its hunters at great speed from Deptford to Greenwich; as a result of this south London sleigh ride, the Royal College of Surgeons acquired the animal's skull. In October 1842 a whale described as a 'fin fish' appeared near Deptford Pier, whereupon five sailors from the Dreadnought

Seamen's Hospital-ship put out in a boat, armed with a 'large bearded spear' and 'commenced the attack upon the monster, which soon showed symptoms of weakness, and threw up large quantities of water from the blowing apertures on its back'. Surrounded by other boats, it was roped out of the water and onto the pier, where the crowds were such that the constabulary were called to restore order. The creature – most likely a minke – was fourteen feet six inches long, with baleen and a white belly. It was subsequently taken by carriage, and several horses, to a butcher's shop on Old King Street, where it was placed on a stand for public display.

It was notable that these whalish strays appeared at precisely the point in London from which their hunters had set out, as if returning to haunt them. In the 1880s a bottlenose whale said to be forty feet long beached off the Woolwich Arsenal. 'It came up the river with the tide, and, when it found itself stranded on the reed bed, blew furiously and turned half-a-dozen somersaults, injuring itself on the stones, and colouring the river with its blood.' The crew of the steam tug *Empress* took a rope to it and towed it off the beach, 'with the intention of consulting with the Thames Conservancy officers as to its disposal'. Most extraordinary of all, at least to modern readers, may be the fact that a dolphin stranded at Battersea Bridge in May 1918 was summarily eaten by the museum's 'distinguished correspondents', and parts served at a banquet at the Mansion House. 'The opinions received afterwards were nearly all favourable, and some of them enthusiastic. It is a fact which deserves to be more widely known, particularly during a period of shortage of meat, that the Cetacea furnish meat of excellent quality and high nutritive value.' Sidney Harmer admitted that 'a certain Cetacean flavour, which is not universally popular, is apt to develop on keeping, but it is possible to remove this to some extent by parboiling . . . it is a fact that there are persons who consider Cetacean meat preferable to all other kinds.'

Even in the late twentieth century, dolphins and porpoises were not rarities in the Thames. In 1961 a sixteen-foot minke was seen diving and surfacing in the river as far as Kew, 'followed by a police launch warning boats to keep clear'. Earlier that day the whale had been found on the river bank, having apparently collided with a boat. Inspectors from the RSPCA, along with police officers and other helpers, had dragged it in a tarpaulin to the water, hoping it would make its way back to the sea, but the animal became caught in reeds by Kew Bridge, and soon after died. This interloper was not so innocent in the minds of the newspapers that reported on it, for twenty-four hours earlier an engineer had drowned when his dinghy overturned at Chiswick, close to where the whale was found; and two boys in another boat were nearly capsized by a 'thrashing whale or porpoise'. The accompanying photograph showed two men standing over the presumed perpetrator, as if to accuse it of these crimes.

The hindsight of history seems to allow such transgressions as naturalists eating their own specimens; but few could have predicted that, in the twenty-first century, a whale would swim under Waterloo Bridge, past Charing Cross – almost under the window where Melville stayed – and past the Palace of Westminster, only to strand itself on the Battersea embankment within the sound of the King's Road.

It was an event that became a kind of global circus entertainment. An animal used only to the booms and clicks of its cousins in the open sea was suddenly subject to the confinement and cacophony of one of the world's largest and noisiest cities. Disorientated and distressed, the northern bottlenose whale moved up and down stream with the tides, its flukes flapping furiously, its curiously baby-like head rising plaintively out of the water while people shouted at it and boats surrounded it and helicopters filled with film crews buzzed overhead, transmitting pictures around the world for fascinated audiences to see. When I watched these scenes again, months later, hindsight served only to make them more poignant in the knowledge of what happened next: a pathetic death, deafened and assailed by traffic, trains, boats and people, frightened by those who sought to save it, starving and therefore suffering terrible thirst, trying futilely to follow a dead-end river to the western ocean.

Inevitably, this visitation was seen as a new omen for the world. A month before, six Arnoux's beaked whales had made an unusual appearance in Cape Town harbour, looking, with their strange, stubby, protruding teeth, their brown skins and mottled, veined markings, like primeval denizens of the deep come to confront the modern world with its sins. Only days before the arrival of the London whale, a fifty-foot-long dead finback was taken from the Baltic at Bremen and driven, with a police escort, to the centre of another capital city, to be laid at the steps of the Japanese Embassy in Berlin as a protest against that nation's continuing actions in the sanctuary of the Southern Ocean. And on the same day that the whale appeared in the Thames, four Cuvier's beaked whales beached in Spain, the victims, as subsequent tests would indicate, of naval sonar exercises.

The London whale was doomed from the moment it entered the estuary, and from which it was scooped up and carried in a procession back towards the sea, watched by news crews and crowds on the Thames bridges. As it lay on its inflated pontoon, the whale's frantic muscular movements began to flag. At seven o'clock that evening, it finally expired, somewhere near Gravesend, two hours from freedom. Borne on its rubber bier, its tearful attendants asked that the cameras be turned off in respect for its passing.

To some these scenes evoked the funeral of Winston Churchill, when the hero's coffin was taken down the river by naval barge, and which I watched on television as a young boy, instructed by my father as to its historic importance. To others, it all seemed a kind of collective madness. This princess of whales – for it was a she – became the subject of national debate and newspaper headlines. There were leader columns on how its treatment was a testament to our humanity; and others that claimed, equally, that its appearance was a reminder of the barbaric practices of the whaling nations. The Victorian press would have reacted in quite the same way: entire tabloid sections, edged in black, appeared to commemorate the whale. Others saw satire in its misadventure: one cartoon showed the animal on a flag-draped catafalque in the manner of a royal lying-in-state – only instead of a quartet of Life Guards with their sabres unsheathed, four photographers stood at each corner with their telescopic lenses downturned. Unbeknown to the artist, his image was an echo of a previous century, when the Royal Aquarium's beluga, another public object of mistaken sex and dislocation, had lain in state in Westminster.

By coincidence, the reading at Mass that Sunday was from the book of Jonah, prompting one clergyman to write from Hull to a national newspaper, noting that the passage was the one in which 'Jonah says Nineveh, the London or New York of his day, will be overthrown in forty days. The people cut consumption by fasting and wearing the simplest possible garments and renounced violence. With the oil running out and global warming beginning to gallop and the continuing hideous aggression of the USA, perhaps the poor creature was giving us a hint.' In fact, as its necropsy revealed, the whale died of dehydration and stress. Months later, Richard Sabin showed me its dorsal fin, preserved in a specimen jar at the Natural History Museum. Wrinkled and greyish black, with the central core of cartilage visible where it was removed, the fin retained its last position, bent on one side, a sign of the trauma its owner suffered in its final days.

(The treatment of the London whale contrasts with that of another bottlenose whale which swam up the Humber in 1938. 'The whale . . . went up and down the river many times between Heap House and Keadby,' wrote the Secretary to the Receiver of Wreck in Hull. 'It grounded continuously, and its struggles caused damage to the river banks, whilst its presence in the river was a constant source of danger to shipping. It was on this account that Starkey decided to shoot it.' The carcase was claimed by the Natural History Museum, although only after querying the butcher's bill from W.A. Hudson in Scunthorpe: 'To: Degutting Whale: £5'.)

Throughout the twentieth century dead whales continued to be a source of fascination. In 1931 an embalmed, sixty-five-ton whale arrived at London Docks, the property of the Pacific Whaling Company and destined for display at a Christmas circus. Housed in a specially built case, it required the world's largest floating crane, the London Mammoth, to transfer it from the ship to road bogies, on which it was taken to the circus, 'the journey to be made by night'. One observer, as a young boy, remembered it with a great stick propping open its mouth, covered in tar to preserve it and smelling like roadworks.

Twenty years later, in 1952, a seventy-foot fin whale caught off Trondheim (after being found by helicopters sent out for the purpose) was preserved on a huge one-hundred-foot-long lorry – also said to be the longest in the world – and was trundled overland through Europe, Africa and Japan, appearing in such unlikely places as Barnsley, Yorkshire, before ending up in exile in Belgium. It was a scenario reminiscent of the Hungarian film, *The Werckmeister Harmonies*, in which a travelling leviathan creates psychic upset in a Cold War-era town and becomes an allegory for totalitarianism – 'Some say it has nothing to do with it, some say it is behind everything' – just as the Czech poet Miroslav Holub imagined,

There is a serious shortage of whales.
And yet, in some towns,
whaling flotillas drive along the streets,
so big that the water is too small for them

while another poet, Kenneth O. Hanson, wrote of a pickled whale carried across Wyoming on a flatcar railway truck, 'shunted to a siding the gray/ beast lay dissolving in chains'. I imagine whales in containers, shipped out in a cetacean pogrom, each in its rusty box on rolling stock. A ferry hits a humpback, a freighter carries a fin whale on its upturned bow, whales slump on sandy beaches.

Ah the world, oh the whale.

Man had developed a new relationship with the whale, although, as ever, it was one predicated on his desires, rather than the rights of the animal. Although the Society for the Prevention of Cruelty to Animals had been founded in England as long ago as 1824, and an animal protection law passed in 1835, it would take a long time for whales to be included in this order. 'Yes; the beasts, the birds, and the fishes all prey upon one another,' wrote a correspondent to *The Times*, who had observed the fate of the London beluga in October 1877,

> and man, whom we believe to be nearest to the Great Creator, preys upon them all. If he wants a sealskin jacket, he kills the seal and takes his skin; if he wants a mutton chop, he kills the sheep and takes his chop; and if he wants a live tiger to stare at, he catches a tiger alive and puts him in a cage; and I am afraid that the sight of the dying throes of the whale will have no more effect upon the feelings of the managers of the Westminster Aquarium than the same sight would have to soften the heart of a North Sea whaler as he drives in his last harpoon, because he wants the oil.

Sentiment still gave way to business. On Christmas Eve 1868, Sven Foyn wrote in his diary: 'I thank Thee, O Lord. Thou alone hast done all.' The Norwegian was giving praise for the grenade harpoon he had just patented; a bomb that would explode in a whale's head. A former seal-hunter, Foyn was 'a most fortunate, religious, and good old man, respected and beloved by all who met him', and the maiden voyage of the *Spes et Fides* – Hope and Confidence – with the eponymous Miencke among its crew, set out equipped with his efficient weapon.

Harpoon guns had been used on whales since the early nineteenth century, but Foyn's holy invention allowed his fellow countrymen to pursue the great rorquals that had been beyond reach of the Starbucks and the Scoresbys: blue whales and fin whales, the largest animals on earth. Now no whale, no matter how fast, could escape; as soon as sighted, it was as good as dead. And a dead whale was a good whale to a Norwegian sailor. Soon the Scandinavians were killing a thousand finbacks a year. Humpbacks, too, began to suffer heavily from the new technological era of steamships and harpoon-launchers.

It was a necessary advance, hence Foyn's earnest prayers: the other cetacean species had simply run out. The sperm whale and right whales were so depleted as to make their pursuit uncommercial; and anyhow, the price of whale oil had plummeted since the introduction of petroleum and gas, and in 1879 the first electric light was switched on. The world looked elsewhere for illumination. The eastern Arctic fisheries were almost finished; when the young Arthur Conan Doyle took passage as a ship's surgeon on the SS *Hope* out of Peterhead in 1880, the vessel returned after a six-month voyage having caught just two whales, and had to rely on seals for profit. Dundee remained an important port, historically prospering by marrying Scottish jute with the whale oil needed to treat it: seven hundred whalemen were still resident in the town in 1883, when a humpback swam up the Tay and, after six weeks' feeding on shoals of herring, was harpooned by a steam launch from the *Polar Star,* and was subsequently embalmed and exhibited in Aberdeen, Glasgow, Liverpool, Manchester and Edinburgh.

In America, the industry experienced a fitful revival with the discovery of the bowheads of the western Arctic; these virginal herds were culled for their huge baleen, used for corsets and hoops to amend the female form. But by the beginning of the twentieth century, whalebone was being supplanted by steel and plastic, and as women emancipated themselves from constricted waists and deformed ribcages, it seemed the whales were about to be set free, too. In 1924 the last whaler sailed out of New Bedford. The trade had long been in decline; Charles Chace, one of few remaining whaling captains, refused to take 'New England' boys (that is, white men) as apprentices, knowing that they would be enrolling in a dying industry. The whaling city had turned to cloth rather than cetaceans. Textile mills lined its river banks, employing labour imported from Lancashire as well as the Azores, and steamships ferried tourists to Nantucket and Martha's Vineyard – prettier places than the dank quays where rotting hulks stood derelict off wharves that still smelt of whale oil.

With the decline of American whaling – just one shore-based whaling station remained operational, in California – European whaling expanded to fill the space. In 1904 the armed steamships of the Norwegians and the British opened up the unplundered Antarctic to fulfil a new use for the whale: in the manufacture of nitro-glycerine. In a new century of war, placid animals supplied the world with the raw material to blow itself up. Fifty thousand whales perished during the two global conflicts – as much victims as those whose death and destruction theirs enabled. The same impulse which allowed slaughter on the Western Front seemed to permit the slaughter on the world's oceans. As Europe suffered losses in their millions, the entire population of humpbacks in the South Atlantic were driven to extinction by 1918. Their oil prevented soldiers from suffering trench foot. In his report on stranded cetaceans of that year, Sidney Harmer noted that 'several of the specimens were accordingly used for the manufacture of glycerine for munitions'. Whales, like men, were fodder for war.

The march of events set in motion by Sven Foyn was unstoppable. Within twenty years of the opening up of the sub-Antarctic whaling grounds, 'the rorquals have declined alarmingly in numbers', one author noted in 1925. That year, the first factory ship, the *Lancing*, was launched in Norway. With these 'seagoing abattoirs', the extermination proceeded, paralleling the most bloody century in human history with the death of one and a half million rorqual whales. It was clear that the slaughter could not continue. 'Whales have been killed on such an extensive scale in the Antarctic regions that, had it not been for the fact that the whaling ground in the Falkland Islands is in British territory and therefore under certain control, the whales would have been killed off as they have been in the Arctic regions,' *The Times* reported in 1926. 'It is feared that unless measures are taken in time the whale will become extinct all over the world.'

As part of that attempt to document and limit catches, and to understand these animals before they entirely disappeared, the Natural History Museum sent its scientists to the southern hemisphere. In 1913 Sir William Allardyce, governor of the Falkland Islands, on whose shores the British whaling stations were sited, realized that the new techniques being used to hunt whales in the Southern Ocean were moving at such a rate that the populations would soon be decimated. The Colonial Office in London agreed to his idea of introducing licences, and to assess sustainability, G.E. Barrett-Hamilton of the Natural History Museum was sent out to investigate. Unfortunately, the scientist died of a heart attack shortly after his arrival in South Georgia, as the indefatigable Percy Stammwitz, who had accompanied him there, had to report.

Stammwitz's other letters were full of life, however; his words portray a scene of almost unbelievable, prelapsarian schools of whales. Writing in the year before war laid waste to Europe, his descriptions commemorated the vast and peaceful plenty that existed before man came south – and which was soon to disappear. 'The Whalers say that the Whales are very plentifull [*sic*] in the Southern Seas,' he wrote, '& can be seen spouting in the thousands, round South Georgia, some of them larger animals reaching one hundred feet in length.' This last phrase was underlined in blue pencil when the letter was received by the curator, Dr T.W. Calman, who added a scribbled note: '*Can we suggest getting one for the NHM*'.

Stammwitz worked tirelessly for the museum, and for the whales, in those years. He was as intrepid as any Edwardian explorer, a specimen hunter in his own right, although the trophies he brought back were destined not for the walls of a stately home, but for the cabinets of the nation's museum. As a young man, he would leave his home in Turnham Green – from where his wife would write anxious letters to Sidney Harmer, asking after her husband's whereabouts and whether she should continue to pay his insurance premiums – travelling to the Shetland Islands to work with the Alexander Whaling Company, reporting that fin, sei and minke whales were there in plenty; they were 'hoping for Humpbacks, too'. There he gathered sometimes dubious information on whale behaviour in the dawn of modern cetology – Gunder Jenssen, manager of the company, replied to one query, 'I never heard of Killers attacking Sperm, as the Sperm are regarded as rather a frightening brute and will go for anything, even sharks' – but also sent back body parts from fins to fœtuses, to the delight of his bosses. Back at the museum, Stammwitz took moulds of whale carcases, creating the models that would hang alongside his greatest achievement, the blue whale.

Percy Stammwitz's annual appraisals, still held in his staff file at the museum's library, are testament to his abilities, not only as Technical Assistant, but as a cetologist in his own right. They list the replica whales – killer, beluga, caaing (or pilot) whale; Commerson's and Heaviside's dolphins, white-beaked dolphin, common porpoise, sei whale, and even a young sperm whale – all recreated in plaster by Stammwitz's expert hands from stranded animals he collected, sometimes in harsh conditions. (After one particularly difficult attempt to recover a sixty-foot sperm whale in Yorkshire – which necessitated a formal request for new boots – it was agreed 'that Mr P. Stammwitz may be allowed six days' special leave in view of his long hours and arduous work at Bridlington'.) Stammwitz's loving fashioning of specimen whales was an intimate tribute to their inherent beauty, one that his young son Stuart would inherit as he assumed the same post at the museum under curators who knew him from his boyhood, noting his 'great mechanical skill' and endearing personality and behaviour when sent on his own collecting missions with the Royal Navy.

As the Scoresbys had seen the rise and falling of eighteenth- and nineteenth-century whaling, so the Stammwitzs' careers mirrored the modern rise and fall of whaling, and the zoologists' deepening concern for the future of the whales themselves. As early as 1885, the museum's first director, William Flower, had made a speech decrying the avarice of whaling in Atlantic and Australian waters. It was the work done in the Southern Ocean by those pioneers that laid the foundations for conservation efforts which would save entire species, at the very moment that they came closest to extinction.

As ever, bureaucracy and finance slowed matters, and it was not until 1925 that the Royal Research Ship *Discovery* – a steam-assisted, three-masted wooden sailing ship first built in Dundee along the lines of a whaler for Scott's Antarctic expedition of 1901, and now refitted at Portsmouth – left for South Georgia, where a laboratory was built next to the Grytviken whaling station. Here the scientists could study whales brought ashore, albeit in hellish circumstances. 'Flesh and guts lay about like small hillocks and blood flowed in rivers,' one researcher wrote, '. . . while clouds of steam from winches and boilers arose as from a giant cauldron.' Four years later, a new ship, *Discovery II*, was built, a 232-foot vessel dedicated – as the memorably named Sir Fortescue Flannery declared at her launching – to collecting data that might bring about an international agreement to restrict hunting in the Antarctic. It would be joined by a newly equipped vessel 'of the whale-catcher type', christened after another famous explorer: the RRS *William Scoresby*.

Restraint in whaling, however, came out of self-interest rather than scientific study. British and Norwegian whalers petitioned the League of Nations – formed to prevent another human Armageddon – to request restrictions on the factory fleets. The need for control became all the more urgent, as Sir Douglas Mawson observed from his Australian perspective, in light of the 'tremendous onslaught' of the 1930–1 whaling season, although another correspondent with an evocative name, Arthur F. Bearpark, wrote from his St James's gentleman's club to point out that both Britain and Norway already had voluntary agreements in place.

In 1935 an international agreement was drafted under the auspices of the League of Nations and ratified by Britain and Norway as the major whaling nations, fitfully united against newcomers in the field. It was soon clear that its precepts were insufficient, and Norway approached Britain to suggest that the scope of the agreement be widened. In May 1936 an International Whaling Conference convened in Oslo, with only two members, Britain and Norway; Germany, under its new leader, refrained from attending officially but sent an observer, saying it wanted 'full liberty of action as being the world's greatest consumer of whale oil', both in margarine, and by the soap firm, Henkel's, which had its own 12,000-ton factory ship. It was not a peaceful meeting. After negotiations described as protracted and interrupted by threats of boycotts by Norway, it was agreed 'to prevent excessive diminution of the whale population by restriction by close season and by limitation of the number of whale catchers used . . . with a whale factory ship'. The season would run from December to March. 'It is hoped that . . . thus a somewhat stormy chapter in the history of modern whaling will be happily brought to a close.'

In an echo of its pragmatic attitude to the abolition of slavery, Britain placed itself at the forefront of these ever more urgent attempts to control whaling while admitting its self-interest – even as other acts of diplomacy tried to stabilize a world moving towards war. In May 1937 an expanded international conference gathered in London, with representatives from South Africa, America, Argentina, Australia, Germany, Ireland, New Zealand and Norway. Mr W.S. Morrison, Minister of Agriculture and Fisheries, told the delegates that 'the blue whale would be exterminated if things went on as they were, and the Antarctic whaling industry would soon cease'. A new convention was announced, banning pelagic whaling for nine months of the year. 'In some areas it is prohibited entirely; some species of whales, whale calves, and females attended by calves are protected absolutely, as are also whales below a certain size; and whaling from land stations' – such as those in the southern hemisphere – 'is to be subject to a close season for six months.'

The conference also hoped that other countries, 'particularly Japan, whose operations are rapidly expanding, will adhere to the present Convention'. Although its coastal settlements had whaled for centuries, and the British ship *Syren* had discovered the prolific whaling grounds off Japan and the Bonin Islands in 1819, it was a visit by Tsar Nicholas II in 1891, who saw great numbers of whales in the Sea of Japan, that prompted modern whaling in Japanese waters. By 1934, using techniques learned from the Norwegians, Japan was making its first whaling voyages into the Southern Ocean. By declining to join the international agreements its industry prospered, and within five years it had six factory ships operating in Antarctic waters.

There was already a degree of equivocation about this east-west split. While they adhered to self-imposed controls – in May 1939 a Norwegian captain was prosecuted for killing a fifty-nine-foot female blue whale, below the limit of seventy feet – Britain and Norway remained responsible for ninety-five per cent of the annual toll of thirty thousand whales, each sending out ten mother ships busy making orphans. The remainder were divided between Germany, Russia, Holland and Japan; America had just one mother ship in an industry it had once dominated, but which Ishmael would have found unrecognizable. There was little that was heroic about this chase, as catcher boats fired on whales from the safe vantage point of prows towering high above the water. In the days of Yankee whaling, at least the whales could fight back; now they no longer stood a chance. A whale once seen was as good as dead.

By the outbreak of the Second World War, huge ships with crews of two hundred and forty were catching five hundred thousand tons of whale a year. As Mary Heaton Vorse wrote in Provincetown: 'the destruction has been so great that the size of the huge monsters is becoming smaller each year, and unless international action is taken the whale will become one of the fabulous monsters of the past.' The whale had become an unwitting symbol of a century of suffering. It was no coincidence that Auden, himself now an exile in America, wrote in his poem, 'Herman Melville', in March 1939, 'Evil is unspectacular and always human'.

The whale had become the enemy by default. Every kind of device was used to kill the animals: exploding harpoons, strychnine, cyanide and curare poisoning (inspired, perhaps, by the Aleutian Islanders who used barbs contaminated with rotten meat to infect a whale with blood poisoning). Even electrocution was attempted: the same method by which the civilized world got rid of its most venal criminals was brought to bear on dumb animals. The hunters came armed with cannon and bomb-lances, ostensibly hastening death, but in practice causing what we can only imagine to be agonizing pain; an apparent indifference to the dignity of animals illustrated by the fact that men on Antarctic whaling stations threw penguins on their fires, using the oily birds as kindling.

As cetacean war was waged from above, using aeroplanes to spot their targets, bombers mistook whales for submarines below, with the inevitable consequences. British and Norwegian ships left the dangerous Atlantic for the Pacific coast of South America; from 1941 to 1943, a Norwegian flotilla working off Peru captured 8,500 sperm whales. These young men working on the whalers were as much a part of the war effort as my mother, busy making machine gun parts in a Southampton factory, or the now elderly Percy Stammwitz, proud to serve in the Home Guard, defending London during the Blitz. War even evoked the whale in an animated propaganda cartoon, in which an insular Britain was menaced by a Nazi whale morphing out of the map of Europe, with Scandinavia as its sinister swastika'd head and the Baltic as its evil-toothed jaw.

As German U-boats extended their operations south of the Equator and the Pacific too became a theatre of war, whaling virtually ceased. Some shore stations still operated in South Africa and Australia, but most whale-catchers were converted to warlike uses, 'and such of the large factory ships – some of them displacing more than 17,000 tons – as escaped destruction are required for more urgent purposes', *The Times* noted under the headline, 'WAR AND THE WHALE'. 'It will be interesting to observe the results of the close season imposed by war,' added the newspaper, which hoped that the rapid decline in numbers noted in the last open season of 1939–40 'will prove to have been but temporary. At the same time,' it admitted, 'the virtual extinction of the Greenland whale, the Pacific grey whale, and the Biscayan and southern "right whales" is a warning against optimism.'

The aftermath of war did not bring peace for the whales any more than it did for their fellow species. Whale oil – and meat – was more valuable than ever as a supplement to rationed diets, and the whaling nations agreed that, in the first year after the war, the season would be extended. In 1945, only months after the cessation of hostilities, the first British whaling steamer to be built since the war sailed from the Tyne for South Georgia in her colours of red, white and blue, with a crew of four hundred on board. 'The *Southern Venturer* is in a hurry . . . The vessel, which has only just been completed, will not reach the whaling waters by the official opening of the season.' Two Norwegian vessels would arrive before her; but so too would two British whalers, converted from captured German ships.

The urgency was to feed a hungry nation. A new technique had been developed of shipping dehydrated whale meat – 'which is said to have high degrees of proteins . . . and of digestibility' – and soon recipes were appearing in the popular press. 'How to Cook Whale Meat – Goulash Recommended' ('Add tomato ketchup to colour . . . Serve with macaroni or dumplings'), while 'Waleburger Steaks' – deliberately misspelt, perhaps to obscure their origin – appeared on the menus of London restaurants. ('After he had eaten his "Waleburger" Mr Lightfoot said he was agreeably surprised by the taste . . . There was no flavour of fish.')

'Whale meat was neither fish nor fowl,' Dr Edith Summerskill of the Ministry of Food admitted, 'but it was now hiding its "Jack

Tar" accent and insisting upon roast beef connexions. As a result all the whale meat obtained was being accepted and sold.' At one shilling and tenpence the pound, it was excellent value, and could be grilled, braised or minced, and served with fried onions, mashed potatoes and Brussels sprouts, although one commentator noted, 'it might be advisable to eat sparingly of it until the digestive system has become more familiar with it.' Meanwhile, in Norway, the Red Cross gave sperm whale teeth to the war disabled for scrimshanding, much as British veterans assembled paper poppies.

Still in wartime mode, Britain applied the lessons of war to whaling. In June 1946 it sent out ships equipped with 'sonic submarine detectors' to find whales, using ultrasonic nets to keep them within range of the boat. Although these techniques were soon found inferior to the animals they mimicked, the renewed industry gathered pace. On 10 May 1948 the whale-ship *Balaena* – and her crew of seventy British and five hundred Norwegians – returned triumphantly to Southampton, having captured three thousand whales, ten per cent of the total catch that season, among them a monster measuring 94 feet and weighing 180 tons.

This massive ship – complete with her own laboratories, blacksmith's shop and hospital – stood alongside the docks I knew as a child, rivalling the ocean liners for size and presence. Her Antarctic cargo – a contrast to the heatwave in which she had arrived – may have been less glamorous than the Hollywood stars those passenger ships carried (Lana Turner was the next celebrity to appear on the quayside), but it was a major contribution to the national economy: 4,500 tons of meat, 163,000 barrels of edible oil (destined to make margarine), 10,000 barrels of sperm oil, 170 tons of meat extract, and a further 3,000 tons of meat for cattle fodder. Set alongside reports of Churchill's exhortations to a

United Europe in the local paper, the *Balaena* and her contents represented hope for a post-war world.

Such resources soon became the cause for resentment, especially as the Americans were aiding the Japanese in their own whaling operations. These were, after all, austere times, and the Allies encouraged the vanquished nation to feed its population fried whale or parboiled blubber as a cheap source of protein. The occupying powers, under General Douglas MacArthur, also helped equip decommissioned naval ships for the purpose; vessels that had fought against the Allies now turned their tonnage towards the whales. They did so against strong opposition from Australia, which complained that the Americans had not consulted it in the matter. It was nervous at the notion of former enemies sailing in its waters, and protested 'on the ground of earlier Japanese violations of international whaling regulations and the inefficiency and wastefulness of Japanese whaling'.

That year, 1948, a Japanese whaling expedition sailed six thousand miles to the Antarctic, carrying a crew of thirteen hundred men, enough to populate a small town (or invade it, as some Antipodeans feared). This modern Armada comprised six catchers, a ten-thousand-ton factory ship, the *Hashidate Maru*, two processing ships to refrigerate its spoils, an oil tanker, and two vessels for cold storage. A Nantucketer would have blinked his eyes in wonder. The ships travelled far apart to avoid collision with each other or with icebergs – using radar to navigate through thick banks of fog – until they came upon their appointed foe: a gigantic blue whale.

A catcher boat was sent ahead, but whenever it had the whale in its sights, the animal sounded. It was two hours before the gunner hit his target. The first deep cut was made then and there, at the geographical point of its demise, for fear of the animal's extraordinary metabolism. Insulated by thick blubber, whales generate tremendous heat – as their condensing blows indicate, akin to great steam engines. If they were to over-exert themselves in pursuit of prey, they could die of heat exhaustion; hence their need to regulate their temperature by cooling their blood in their flukes and fins. A whale killed in the Southern Ocean was immediately slit open from throat to tail, allowing cold water to flush through it, lest its internal heat cause its very bones to combust, leaving its hunters with a 'burnt whale', burning on its own oil like a giant candle, just as its brethren once burned to light the world.

Towed back tail first and up through a ferry-like skidway in the ship's stern where eighty men worked for four hours to butcher it, the blue whale was one of the largest ever caught. It weighed 300,000 pounds, although they only knew this because they were able to slice it into pieces and place it on the ship's scales. The tongue alone weighed three tons; the heart was as big as a car, and the arteries wide enough for a man to swim in. All was now so much offal.

And all this was accomplished in an atmosphere of outright hilarity. 'Workmen laughed and leaped aboard loins that were skidding toward the loading chute,' observed Lieutenant-Colonel Waldon C. Winston, an American officer accompanying the fleet. 'Others there started a shanty. Over and over, they filled the box on the small platform scales, then emptied the contents down the loading chute.' They might as well have been on a Detroit production line.

Below decks were steel boilers where the blubber was reduced to oil which was then stored in huge tanks. Nothing was wasted. A process had been devised to suck vitamin-rich oil from the whale's liver. This one animal yielded 133 barrels of oil and sixty tons of meat valued at $28,000. This process went on, day by day, month by month, year by year, in waters so far from land that wounded men often died, there being no hospital to which they could be taken.

Here, out of sight, off shores belonging to no one, no one was responsible. Yet as the ships canned their whale meat, official observers looked on, and biologists sought to learn about living whales by examining dead ones. It was a uniquely mad situation, belied by its own legitimacy. Although regulations stated that mother and calf pairs were not to be targeted – any gunner who shot them had his pay deducted accordingly – pregnant animals were taken. These were the hardest to kill; one blue whale mother took nine harpoons and five hours to die.

In ancient Japan, Buddhists had honoured these unborn cetaceans, erecting stone tombs for them facing the sea, so that at least in death they could see the home they had been deprived of in life. American scientists working on the ships had other plans. One who found a five-inch sperm whale fœtus had it packed in ice, and back in port at his hotel used a mixture of vodka and shaving lotion to preserve it overnight. The next morning he dissected the specimen. It had the rudimentary features of the animals that became whales: with its pig-like snout, nostrils positioned at the front (before they migrated up the head), its protruding ears and genitals, and its hand-like flippers and residual whiskers, it was as if this whale-in-being might yet become some other creature entirely.

Only in death could man see whales in such detail; only on these mother ships were the massive animals seen to be colonies in their own right, living cities crawling with whale lice and studded with barnacles which finally loosened their grip as the blankets of blubber were cut away, the hard shells popping out of the epidermis and clattering to the deck. The whale's interior played host to other parasites: the nematode worms that colonized its guts (intestines which, to scientists' amazement, unravelled for a quarter of a mile). The *Hashidate Maru* merely minced up these worms along with the rest of the meat. Of more concern were the levels of radioactivity to be found in whale flesh, fallout from the devices that had exploded over Hiroshima and Nagasaki. But by then, every man, woman and child on the planet was absorbing strontium-90 into their bones from those explosions, a legacy to be passed on for generations to come.

In iceberg-blocked waters, serried ranks of rorquals lay belly up like gutted herring, side by side while sea birds fluttered about them like feathered stars. They were captive whales, ready for rendition. A factory fleet could cull seventy animals in one day, using weapons that resembled space-age missiles, flanged and fluked to implode in giant crania. Three hundred and sixty thousand blue whales died in this manner in the twentieth century, reducing their population to one thousand. By the 1960s the blue whale was, to all intents and purposes, commercially extinct.

Captain Ahab
The White Whale nearly destroyed him—and no one would rest till the White Whale was dead!
Queequeg
Cannibal-turned-harpooner—his fierce strength as strange as his mystic tattoo!
Ishmael
Handsome, romantic—the look of a poet, the courage of a giant!
Father Mapple
The blood-and-thunder preacher who roared his warnings to the waiting women!
Starbuck
He bet his fears and his faith against the fury of Captain Ahab!
Tashtego
The silent Indian—when hunting buffalo became too tame, he went after Moby Dick!
There Is No Might Like
The Might of Moby Dick!
WARNER BROS. PRESENT
GREGORY PECK
RICHARD BASEHART
LEO GENN
IN THE
JOHN HUSTON
PRODUCTION OF
MOBY DICK
HERMAN MELVILLE
AND ORSON WELLES as 'Father Mapple'
FILMED WITH A SPECTACULAR NEW DEVELOPMENT IN COLOR BY
TECHNICOLOR
A MOULIN PICTURE
DIRECTED BY JOHN HUSTON
PRESENTED BY WARNER BROS.

XII

A Cold War for the Whale

You have become like us,
Disgraced and mortal.

Stanley Kunitz, 'The Wellfleet Whale'

For his 1954 film of *Moby-Dick*, made in Britain and Ireland rather than New England, John Huston requisitioned an 1870 schooner which had recently done duty as the *Hispanola* in Walt Disney's *Treasure Island.* She was fitted out at St Andrew's Dock, Hull, where chandlers contributed original harpoons, found in their loft. This movie *Pequod* was then sailed to the west coast of Ireland, where the director chose to shoot only on overcast days to give his film a gloomy look.

I remember watching Huston's film as a young boy; it seemed rather wordy and dull to me. Our old-fashioned, veneered black and white television, with its grainy 405 lines, did little to convey the subtle effect that the cinematographer, Oswald Morris, had devised to emulate nineteenth-century whaling scenes, combining two sets of negatives – one monochrome, the other Technicolor – to suggest 'that this story was filmed in 1843 when it was supposed to have taken place'. I did not appreciate the deftness of Ray Bradbury's screenplay, for which he read the book nine times and wrote fifteen hundred pages of script to reach a final one hundred and fifty: 'I found myself plagued with a vast depression,' said Bradbury. 'I felt I had the weight, the burden of Melville on my back.' Any analogies between the nineteenth century's thirst for whale oil and the post-war desire for petroleum escaped me, too.

Nor was I impressed by Orson Welles's bravura cameo role as Father Mapple, performing histrionically from his pulpit-prow – in Shepperton. (Welles would stage his own version of *Moby-Dick* in 1955 – claiming it was the best thing he had ever done – at the Hackney Empire, in London's East End.) I may have recognized Richard Basehart as Ishmael, but only because he played the submarine captain in *A Voyage to the Bottom of the Sea*, battling a giant squid; as with Disney's *Twenty Thousand Leagues Under the Sea*, both whale and squid were Cold War monsters, subaquatic versions of science fiction aliens, the threat the world faced from within. And although the sight of another savage other, Queequeg and his tattooed face and his long red drawers, was terrifying enough, when a whale finally appeared on screen, it was difficult to tell if it were alive or not – not least because Huston had recreated Moby Dick as a life-size model. (At one point during the filming, part of the 'White Whale' broke loose while being towed off Fishguard in rough seas, causing coastguards to alert shipping to a 'possible hazard to navigation', and the Royal Air Force to send out a flying-boat in search of the errant prop.)

Roped to this ersatz whale, Gregory Peck nearly drowned as Huston insisted on take after take of Ahab's final moments. But it is only now, watching the movie again, that I see something shockingly real in these scenes. Intercut with sequences acted out in a studio tank – betrayed by the wrong-sized waves and an atomically lurid, back-projected sky which turns Gregory Peck's Ahab into a kind of pantomime demon king – Huston inserts footage of sperm whales being hunted off Madeira. Here his film comes closest to the truth, in the mortal spout of dying whales, the gushing crimson fountains. It is an unforgettable, Hemingway-like gesture; only instead of a dying bull, it is the world's greatest predator that perishes, publicly, as advertised, on screen.

In 1958, the year in which I was born, Ernest Hemingway told the *Paris Review* that he had hunted a school of fifty sperm whales, and harpooned one 'nearly sixty feet in length and lost him'. His was a forlorn boast of a heroic American past. Whaling was now the province of other countries, and their efforts would do far more to bring whales to the brink than the Yankee fleets ever did. In fact, it was within my lifetime that whaling reached its all-time peak. In 1951 alone – one hundred years after Melville's book appeared – more whales were killed worldwide than New Bedford's whale-ships took in a century and a half of whaling.

In my *Illustrated Animal Encylopædia*, edited by curators from the American Museum of Natural History and illustrated with photographs of the museum's dusty dioramas – although not, I'm glad to say, with its positively horrifying set-piece of a life-size sperm whale doing battle with a giant squid – the limits of 1950s cetology were acknowledged. As if in response to Ishmael's question, 'Does the Whale Diminish?', the authors issued a tardy reply. 'We cannot hope for much success until we know more about these deep-sea mammals. We are seriously endeavouring to get this information.'

The book bears witness to a pre-ecological age. A section entitled 'MOST IMPORTANT PRODUCT FROM THE WHALE' states that 'One recent whaling season in the Antarctic produced 2,158,173 barrels of oil', but under another headline, 'THE WHALE IN DANGER', it reports that 'whalers took 6,158 blue whales, 17,989 finback whales, 2,108 humpback whales, and 2,566 sperm whales in a single season . . . This does not include 2,459 whales taken by the Russians.'

It is salutary to see how sharply the figures escalate throughout the twentieth century. In 1910, 1,303 fin whales and 43 sperm whales were taken; in 1958, the totals stood at 32,587 fin and 21,846 sperm whales. It was a momentum exacerbated by politics. From 1951 to 1970, the Soviet Union increased its catches outside international agreements, taking more than three thousand southern right whales, although only four were reported to the International Whaling Commission. First convened by President Truman in Washington, DC, in 1946, the IWC – whose headquarters were based in Cambridge, England – introduced successive steps to limit whaling further, but commercial pressures and unsustainable quotas overtook good intentions.

Humpbacks were particular victims of this slaughter. The Russians claimed to have taken just over two thousand animals, but the figures show that they killed more than forty-eight thousand. Young whales, mothers and calves, protected species were taken indiscriminately, and the figures falsified. Sperm whales, too, suffered badly. At the turn of the century they had enjoyed an illusory reprieve as the newly mechanized fleets began to pursue the rorquals, but after the Second World War, with baleen whale populations rapidly shrinking, the harpoons were aimed again at the cachalots, whose numbers had just begun to recover.

By the 1950s, at the height of a new antagonism between east and west, an average of twenty-five thousand sperm whales were dying each year, ending up as vitamin supplements or animal feed. 'Boiled sperm whale flesh can be used for feeding fur-bearing animals,' noted one Russian scientist, Alexander Berzin, a Soviet-era Beale whose book was illustrated with indistinct images of whale pathology and dissection. His countrymen also used the tendons in the whales' heads to make glue, and in 1956 alone, 980 tons of whale hide were processed in a single Russian factory, tanned and dyed and destined to make soles for shoes. Men walked on whales.

The Cold War was taken to the whales in their ocean fastness. North Atlantic right whales, protected since 1935, were reduced to one hundred animals by the USSR, which also killed 372 of the even rarer North Pacific right whales. Southern right whales had already reached a low point of tens off the coast of apartheid South Africa, while Arctic bowheads suffered similarly under the disunited nations.

The reason for this renewed interest was, of course, financial. Whaling was rapidly becoming the province of new multinational companies. By 1957 whale oil was fetching £90 a ton; in Oslo that year Unilever acquired 125,000 tons of the stuff, from Norwegian, Japanese and British manned ships, although, when asked, the company declined to comment on its purchase. A few years later, it was estimated that whales were worth £50 million a year to the global economy. Helicopters were being used to spot whales in the Southern Ocean, where one whale-ship received a regal gam when the Duke of Edinburgh boarded the *Southern Harvester* from the royal yacht *Britannia*, the princely person being hoisted across in a basket slung from the masthead, while a fifty-foot sperm whale provided the buffer between the two vessels. (Later, the Duke was heard to remark in a television interview that 'A whale has an odour peculiar to itself.')

The tenth meeting of the International Whaling Commission at The Hague in 1958 implemented new restrictions, extending the existing prohibition on killing humpbacks in the North Atlantic and in part of the Arctic, and limiting their hunting in the Antarctic. Such limits meant nothing to those who had not signed up to the organization. 'The whaling industry lives with a recurring nightmare: the extinction of the whale,' *The Times* stated in a forthright editorial in January 1959, foreseeing 'a massacre in the Antarctic next season'. It demanded neutral observers and a ban on the building of new whale-ships without consultation. 'Britain's role as peacemaker is a laudable one. It can, however, hardly be pursued to the detriment of British whaling, and cannot be pursued without the cooperation of others.' As another scientist pointed out, 'conservation had failed mainly because whales belonged to no one and it was no one's direct interest to look after them.'

While the IWC investigated more humane ways of killing whales, Holland and Norway – two of the so-called 'big five' of the whaling nations (the others being Britain, Japan and Russia) – announced their decision to withdraw from the convention 'because it has proved impossible for the two countries to obtain reasonable whaling quotas'. Even as the western nations squabbled, Japan was increasing its fleet. By 1963, headlines were announcing, 'TASTE FOR WHALE MEAT BOOSTS INDUSTRY' – a reference to the fact that for the Japanese, whale oil was secondary to its meat – and noting that the nation had recently acquired the *Southern Harvester*, the same ship visited by the Queen's husband. There was a certain bias to such reports – 'There is a mechanical ruthlessness about Japanese whaling methods which makes the whalers of a few years ago look like amateur adventurers' – which was another legacy of war.

The same article added that a 'state of piracy' was 'gradually emptying the whaling grounds'. Whaling was a free-for-all, and one of the worst offenders was the Greek shipping magnate, Aristotle Onassis, future husband of the former First Lady. His vessels were purposely registered in Honduras and Panama, countries beyond the IWC's membership, and plundered protected waters, taking whatever whales they met, 'be they endangered species or newborns'. Only when Norway publicized his actions – and after the Peruvian navy and air force had opened fire on his ships for hunting whales within their territorial waters – was Onassis forced to stop his slaughter, finding it more financially viable to sell his fleet to the Japanese.

All this was accomplished despite – or perhaps because of – quotas imposed by the International Whaling Commission. The Antarctic catch for 1967–8, for example, was set at thirty-two thousand 'Blue Whale Units'. The world's largest animal was reduced to a mathematical quantity, and its ancient population as 'stocks' in bureaucratic equations. It was a terrible arithmetic:

1 BLUE WHALE UNIT = 2 FINBACKS,
OR 2½ HUMPBACKS, 6 SEI WHALES.

Not only was the average size of whales in the catch declining, 'which points suspiciously to overkilling', as one scientist noted, but 'the CDW – take per catcher's day's work – which is a measure of the effort required to take a whale, is also steadily declining, which tells us what we already know, that the whales are disappearing'. An awful possibility led another marine biologist to wonder, 'What will be next? Will the orbiting satellite speak through space to tell the hunter where to find the last whale?'

The whales could not win. As the rorquals diminished in the Antarctic, the whaling nations turned back to sperm whales. Many thousands were caught by the fleets on their way to the Southern Ocean, in warmer waters where females and breeding stocks were found. During its London meeting in 1965, the IWC discovered 'massive evidence' to show that regulations about the size of sperm whales that could be taken were being comprehensively broken. As a result, the commission banned sperm whaling between latitudes of 40° north and south. That year the killing reached its historical crescendo, with the death of 72,471 whales.

One of the last whaling ports was Dundee, sending ships to waters that, twenty years later, would witness Britain's last colonial war, and where they now lie as rusting wrecks in the rocky harbours of South Georgia and the Falklands. Some of the men who worked on the ships are still alive, and describe their work in these open-air abattoirs as an inferno. The remembered noise, the smell, the sights repulse their memories, retrospectively. If the whales had been able to scream, they say, no one would have been able to bear their work. Instead, the whales were rendered dumb in the face of destruction, as if they agreed not to protest against their abuse, the more to shame their persecutors.

Rejoice

I cannot claim immunity. As I walked home from school through wet autumn leaves to find my mother drying clothes by the fire, Southampton factories were processing whale-oil margarine which sat in yellowy blocks in our fridge, while my cheeks were brushed with whale fat, for 'women will be interested to learn it goes into the making of their cosmetics', as my encyclopædia informed me.

The lingering smell of whale.

While I read illicit American comics under my bedclothes, fantasizing about a world of sleek-suited superheroes, new processes – sulphurization, saponification, distillation – extended and rationalized the use of whales in lubricants, paint, varnish, ink, detergent, leather and food: hydrogenation made whale oil palatable, sanitizing its taste. Efficiency ruled, in place of the early whalers' waste. Whale liver yielded vitamin A, and whale glands were used to make insulin for diabetics and corticotrophin to treat arthritis. Nineteenth-century trains had run on whale oil; now streamlined cars with sleek chrome fins used brake fluid made from the same stuff. Victorian New Englanders had relished doughnuts fried in whale oil; now children with crew-cuts and stripy T-shirts licked ice cream made from it. Their bright shiny faces were washed with whale soap, and having tied their shoelaces of whale skin, they marched off to school, past gardens nurtured on whale fertilizer, to draw with whale crayons while Mum sewed their clothes on a machine lubricated with whale oil, and fed the family cat on whale meat. In her office, big sister transcribed memos on typewriter ribbon charged with whale ink, pausing to apply her whale lipstick. Later that afternoon, she would play a game of tennis with a whale-strung racquet. Back home, Daddy lined up the family to take their photograph on film glazed with whale gelatine.

Whales imprinted with the image of the age.

It was not until 1973, when I was a teenager, that Britain began to ban whale products. Even then, it allowed exceptions such as sperm whale oil, used as engine lubricant, and spermaceti wax, for softening leather – of which it still imported a total of two thousand tons a month – along with other products 'incorporated abroad into manufactured goods. Sperm whales had not been overexploited,' said the Minister of State for Agriculture, Fisheries and Food, 'but baleen whales had.' The Pet Food Manufacturers' Association – which used ninety-five per cent of all imported whale meat to give its dog and cat foods '"chunky" appeal', announced that it would accept its last consignment that November.

Whales may no longer have lit the world, but time still ran on their oil. Watchmakers used the superior lubricant, prized in polar latitudes for its ability to allow chronometers to function in freezing temperatures (thereby allowing the hunting of more whales by Antarctic fleets). As the giant astronomical clock of Strasbourg Cathedral ceremonially tolled European hours, it did so lubricated by the products of William Nye's Oil Works, New Bedford.

And while whale-oiled clocks ticked, the mythic beast acquired a new meaning in the half-life of the nuclear era. In the late 1940s the American artist Gilbert Wilson became obsessed with Melville's novel, as well as with modern science. In the *Bulletin of the Atomic Scientist*, he wrote of *Moby-Dick* that 'no tragedy in world literature succeeds quite as powerfully or as clearly in pointing up the mortal errors of hate and domination'. Wilson even suggested to Shostakovich that they should create an opera of *Moby-Dick* as 'a catalyst for helping to dissolve American and Soviet cold war dissension and to restore world peace'.

In Wilson's own dystopian imagination, the White Whale became an augury of atomic conflict, and Ahab's 'insane pursuit of Moby Dick into the Sea of Japan' analogous to America's 'atrocious nuclear experiments and explosions in the same area'. Similarly, in his critical work, *The Trying-Out of Moby-Dick*, published in 1949, Howard P. Vincent considered that Moby Dick was 'ubiquitous in time and place. Yesterday he sank the *Pequod*; within the past two years he has breached five times; from a New Mexico desert, over Hiroshima and Nagasaki, and most recently, at Bikini atoll.'

A cloud like a whale.

A generation earlier, D.H. Lawrence, writing in Lobo, New Mexico, had seen in Melville's book the 'doom of our white day . . . And the *Pequod* is the ship of the white American soul.' In 1952 the Trinidadian writer C.L.R. James was detained on Ellis Island. Exiled within sight of Manhattan's towers, in a dull brick block next to Liberty Island, James composed his critique of *Moby-Dick*, comparing Ahab with modern dictators. In James's essay, written in the running shadow of the nuclear race, Ahab's *Pequod* became a weapon of mass destruction. 'He has at his sole command a whaling-vessel which is one of the most highly developed technological structures of the day. He has catalogued in his brain all the scientific knowledge of navigation accumulated over the centuries. This is one reason why he is so deadly a menace.' Such potential imagery could be turned against the west, too. Twenty years later, the anti-capitalist terrorists of the Baader-Meinhoff gang, imprisoned for pursuing their own war on imperialism, assumed code names from *Moby-Dick* (with Baader himself as Ahab) – seeing the monster of Melville's myth, as much as Hobbes' state, as their target. Even now, Ahab's crazed pursuit remains the currency of political satire as world leaders are likened to Melville's dæmonic captain in the 'war on terror'.

> He who fights with monsters should be careful lest he thereby become a monster. And if thou gaze long into an abyss, the abyss will also gaze into thee.
>
> Friedrich Nietzsche, *Beyond Good and Evil*

By the 1960s cetaceans were being bodily enlisted into the military. The US Navy instituted its Marine Mammal Program, teaching bottlenose dolphins and beluga whales to identify mines and even act as underwater sentries. Dolphins served in Vietnam where it was rumoured that they were trained as assassins, using needles fitted to padded nose-cones and cartridges of carbon dioxide to deliver body-imploding doses of gas to Vietcong divers attacking American ships. They still play a part in warfare, deployed in the last Gulf War to clear mines from the port of Umm Qasr using cameras strapped to their pectoral fins. To some, such conscription was the ultimate perversion of the relationship between man and whale.

Human technology was catching up with its cetacean equivalent as machines mimicked whales themselves. In one experiment, a whale's skin was replicated in rubber on a submarine's hull, where it was found to reduce turbulence and drag; as a result, protruding parts such as radar dishes and conning towers were sheathed in rubber. This may have been the reason why one submarine was found with the sucker-marks of a giant squid. It had, it seems, been mistaken for a whale.

The development of marine acoustics during the Second World War had alerted the military to the sounds made by whales (which whalers had once mistaken for ghosts in the ocean as they heard them through the hulls of their ships). As the undersea world which everyone had assumed to be a silent place was discovered to be alive with noise, it was suggested that submarines could be disguised as whales by playing their recorded sounds. A century before, slave ships operated under the guise of whalers; now nuclear submarines sought the same deceit. Cetacean technology allowed man to invade the whales' world, in the process creating sounds that would prove fatal for them.

As below, so above. While robo-whale submarines imitated them in the depths, lubricated by sperm oil which would not freeze at great depths and echoing with the ping of cetacean-inspired sonar, whales enabled the exploration of another extreme environment, as NASA used sperm oil for its delicate instruments and rocket engines, sending a trace of whale genes into outer space. Two centuries before, whales had sparked rivalry between Atlantic states; now they were part of the space race. One scientist who sailed with whaling fleets in the 1950s and 1960s told me that it was only when it had a lifetime's supply of oil – I imagine marked barrels sitting in some secret cellar – that America lobbied for a ban on hunting sperm whales (despite the protests of the Pentagon). The fact that the US evolved chemical substitutes for other military uses of whale oil, while the USSR relied on sperm oil for its tanks and missiles, further fuelled the brinkmanship. Even now, space agencies in Europe and America still use whale oil for roving vehicles on the moon and Mars; and as you read this, the Hubble space telescope is wheeling around the earth on spermaceti, seeing six billion years into the past, while the *Voyager* probe spins into infinity playing the song of the humpback to greet any friendly aliens – who may well wonder at our treatment of the species with which we share our planet.

* * *

To the medieval world, which believed the earth to be flat, and monsters to lie in the oceans beyond their illuminated maps, the whale was a scaleless, naked fish – a convenient confusion that allowed its flesh to be eaten by monks on fast days – just as puffins were thought to be half bird, half fish, and geese were believed to be born of barnacles. Despite Aristotle's investigations in the fourth century BC, when he concluded that whales were mammals, it was not until 1773 that Linnæus classified them as such.

Yet the confusion continued. Nineteenth-century whalers called their quarry fish, a wilful insistence that Ishmael mischievously maintained. Perhaps it was a subconscious evasion, for the hunters knew perfectly well, as they butchered their prey, that the physiology they found within was that of a creature more like themselves than a haddock or a cod. And although twentieth-century hunting revealed much of what we know about the blue whale, for instance, even here there was deception. Dimensions were overestimated because bodies stretched as they were pulled out of the water. The only way of weighing these huge carcases was to chop them up into chunks, so vague guesses were made to compensate for the tons of blood lost before the flesh met the scales. Since whales possess proportionately two-thirds more blood than humans – the better to store oxygen on their time spent in the relative safety of the depths – this was an additionally inexact technique.

It was also a self-interested investigation carried out to ascertain profitability, although many scientists, who realized the likely fate of the whale, had another agenda. In the mid-1930s a scheme began from RRS *William Scoresby* to tag whales. Steel darts were shot into whales and retrieved when the animals were caught; whalers were offered a £1 reward to return them to the

You might think that where a whale goes in its spare time is nobody's business but its own. We thought so, too, but we were wrong. Some scientific people were intensely interested in the subject, and decided that in future whales must wear identification discs. But whales are unreasonable creatures, and move about a good deal in salt water, which in time corrodes almost anything but the whale itself. So the scientists came to us. Why to us? Well, we make a stainless steel tube which is quite blasé about salt water. And from this stainless tube, "darts" are made, which are dated and numbered and fired into the blubber. When the whale is eventually caught, the darts tell the story.

Simple, isn't it? But it's simply nothing to some of the other clever things we do with stainless steel tubes. We have helped to solve problems of corrosion in many businesses. Might we not be able to solve yours too, with Stainless Steel Tubes?

STAINLESS STEEL TUBES

BY ACCLES & POLLOCK LTD · OLDBURY · BIRMINGHAM

Colonial Office, with a note of the time and place of death. The data was then analysed 'to gather information not only on the migrations of the whales, but on the question of whether a whale returns to the same ground in the South year after year'. In 1936, eight hundred whales were thus marked and numbered. The cumulative results were astounding: one blue whale was found to have travelled nearly two thousand miles in less than fifty days.

Cetacean research remained invasive. In 1956 a project was devised by a Boston heart specialist, Dr Paul D. White – famed for attending President Eisenhower – to record a whale's heartbeats on an electrocardiogram. This involved shooting a harpoon that contained the metal contact usually placed on a human patient's chest into the animal; although, as one newspaper noted, 'there is no indication so far of how the whale may hope to shed its burden when the interests of science have been served.' 'It can hardly be said that the animal was benefited by such a diagnosis,' my encylopædia added. Dr White's patient – or victim – was a fifty-foot grey whale. Having already experimented with a right whale, White discovered that the whale's heart beat like that of a human. In the light of such knowledge, attempts were made to limit the animals' suffering: the British experimented with a new electric harpoon for humanitarian reasons, but it did not prove successful.

As I was growing up, watching captive dolphins perform in Brighton's underground aquarium built into the promenade – its yellowy-lit, car-park interior echoing with their clicks and all the more forlorn for its seaside setting – attitudes to whales were changing. They had done so drastically since my parents' generation. In the 1920s the porpoise was so abundant in the River Tyne that salmon fishermen were urging 'that steps should be taken for its destruction'. By the 1960s *The Times* was running a headline, 'BRITISH COLUMBIA STIRRED BY DEATH OF WHALE', noting that the demise of an orca had become national news in Canada.

The animal – dubbed 'Moby Doll' – had been harpooned by the curator of the Vancouver Aquarium, where it was destined to be used as a model for a plaster replica. But the whale survived its wound and instead followed the boat. The public reaction was, perhaps, the first real indication of a new attitude towards whales. Although having extracted the harpoon, the animal was kept alive with horses' hearts, flounders injected with blood, and baby seals ('There were immediate protests from humanitarians') and when it was found dead on the bottom of its tank, it was discovered 'that Moby was no Doll. Moby was a bull.' The following year, another orca was captured in the same seas. Named Namu, it was towed in a gill net trap four hundred miles south to its new home, the Seattle Aquarium, only for its captors to discover, two hours out of Port Hardy, that their charge was suddenly surrounded by forty killer whales 'who seem determined to free him'. It was reported that Namu's family had come to visit him, and that he could recognize them by their markings and scars.

Through individual animals such as Namu, whales became emblems of a new age. In the 1960s the American scientist Dr John C. Lilly made controversial claims about cetacean intelligence which led him to an equally extraordinary declaration. 'We need a new ethic,' he wrote,

> new laws based on those ethics which punish human beings for encroachment on the life-styles and the territory of other species with brains comparable to and larger than ours. We

> need modifications of our laws so that the Cetacea can no longer become the property of individuals, corporations, or governments. Even as respect for human individuals is growing in our law, so must the respect for individual whales, dolphins, and porpoises.

Dr Lilly, whose plea echoed Henry Beston's of the 1920s as well as the new age ethos of his own time, went so far as to state that dolphins were 'probably quite as intelligent as man but in a strange and alien way, as a consequence of their life in the sea', and that whales had 'a complex inner reality or mental life'. However, his fellow scientists regarded his investigations with a degree of scepticism, not least because Dr Lilly was also experimenting with LSD in his researches into human consciousness. He also went on to advise on the making of the 1973 film, *Flipper.*

With growing awareness of the whales' plight and the apparent ineptitude of the International Whaling Commission, conservationists began to insist that they were entitled to comment 'on the workings of a group that seemed to have decided that the world's whales were theirs to parcel out'. Yet by the time the horror of whaling truly came to the public's attention – their eyes opened by organizations such as Greenpeace and Friends of the Earth, and acrimonious protests such as the throwing of blood over Japanese delegates to the IWC as well as direct action on the ocean itself – it was too late. The cetacean population of the world had been hunted, harpooned, blown up, butchered, ground down and consumed in a manner unrivalled by any other exploitation of the earth's living resources.

The industry itself had long since ground to a halt by the time the eco-warriors had won their battle; a truly Pyrrhic victory, despite the piecemeal protection of the species: in 1966 all humpback whaling was banned; in 1976 fin whales, and in 1978 sei whales were similarly protected. In the last decades of unrestricted whaling, the Russians and Japanese – who had been forced to turn on the smallest rorqual, the minke whale – resumed sperm whale hunting in the North Pacific, where, from 1964 to 1974, they managed to kill a quarter of a million animals. It was as if, in advance of the end they knew must come, they exerted themselves all the more in the effort.

It was in 1982, in the unlikely setting of the Metropole Hotel in Brighton – a few hundred yards from the town's Aquarium and its performing dolphins – that the IWC instituted its worldwide moratorium, delayed to allow the whaling nations time to comply. Yet the whale remained a victim of international politics, for all its statelessness: threatened by deafening noise pollution which causes ear damage, fatally undermining a sperm whale's most important sense; infected by chemical pollution which generations pass on through mothers' milk to their calves; snared by fishing lines in a frantic, drowning death as bycatch kills three hundred thousand cetaceans every year.

Whales swallow plastic debris by mistake; the thinning ozone layer induces skin cancer; warming oceans have sent their food sources in retreat; climate change has overtaken their ancient knowledge of their environment and its resources. All the while, they move from territory to territory, in and out of legislative areas and through high seas beyond any conserving principle or responsibility, yet forever subject to human activity, wherever they go (even as my own transatlantic movements scar the sky, signwriting my environmental sins in the air).

There is no escape. Sometimes it seems whales are almost pathetically condemned to victimhood. Once they were scared into submission as the harpoon was applied; fed the furnaces with their own scraps; even supplied their own oil to clean up afterwards, as if to apologise for the mess. Now they are early warning systems of ecological attack, as if their own sonar were detecting destruction. Ishmael's cetacean utopia looks further away than ever. Stressed and pressurized by our relentless encroaching on its environment, the whale can ill afford to suffer further sustained periods of active pursuit, although that is exactly what it faces. Since 1987, when the international moratorium eventually took effect (with exceptions for aboriginal subsistence hunts for Inuit populations in Greenland, Russia and Alaska, Makah native peoples in Washington state, and the Caribbean residents of St Vincent and the Grenadines), an estimated twenty-five thousand great whales have died. Japan alone has killed, under its Antarctic Research Programme, JARPA, and its North Pacific equivalent, JARPN, 7,900 minke whales, 243 Bryde's whales and 140 sei whales, as well as 38 sperm whales, which it resumed hunting in 2000. In 2006 JARPA II took 1,073 minke whales – known to their hunters as 'cockroaches of the sea' – and added fifty fin whales to the tally. Each year Japan kills twenty thousand smaller whales, dolphins and porpoises not covered by the moratorium.

Although this meat appears on open sale in Japanese markets, conservationists claim that much of it is stockpiled due to diminishing public taste, or ends up as pet food. Some is sold as meat from baleen whales, which are less susceptible than toothed whales to contamination in the food chain, although in fact it is from odontocetes. The value of JARPA's research is challenged by other scientists, who consider that it has produced no data that could not have been gathered from non-lethal methods. As with Norway and Iceland, which hunt minke whales quite openly, other, cultural motives are at work. Like Europe, Japan claims whaling as part of its heritage for thousands of years – historically encouraged by rulers who forbade the eating of land animals.

Japan also points out that aboriginal whale hunts take place in American waters every year; what is the difference between that and their own claim to cultural precedence in coastal towns? Humpbacks are still hunted on the Caribbean island of Bequia, to techniques learned by a Bequian fisherman engaged by a Provincetown whaler in the 1870s. In 1977 it was said that the United States was 'embarrassed' by the continuing Inuit hunt of bowheads, of which fewer than two thousand survived. 'They push vigorously for smaller quotas every year, and nag the Japanese and Russians . . . remorselessly. Unfortunately, one of the whales closest to extinction, the bowhead, is hunted exclusively by Americans.' While the Inuit had long traditions of sustenance and religion associated with the whales, now that they were 'rich enough in oil money to buy motor boats, powerful rifles and explosive harpoons', whale hunting 'has ceased being a ritual or a means of survival, and has become a sport'.

Since Japan was encouraged and even assisted in post-war whaling by the west – whale meat was served in school lunches until the 1970s – it irks to be lectured on the subject. 'It's not because Japanese want to eat whale meat,' Ayako Okubo told the *New York Times*. 'It's because they don't like being told not to eat it by foreigners.' Some contest that it was actually America's over-use of pressure on the Japanese – and the moral weight of the environmental lobby – that pushed Japan into its intransigent position. Indeed, although America was highly vocal in the anti-whaling campaign of the 1970s (presenting a proposal to a 1972

United Nations conference on the environment to ban all whaling for ten years), things might have been very different if, like Russia, Norway and Japan, the US had maintained a whaling presence in the post-war years. If its industry had not failed in the late nineteenth century, there might not have been the political impetus to ban international whaling. Perhaps this is the true legacy of *Moby-Dick*.

It is true that stocks are recovering from the nadir of the mid-twentieth century. Numbers of humpback and minke whales are increasing in southern and northern oceans, and the southern right whale, *Eubalæna australis*, is breeding successfully off the coasts of South Africa and South America, raising hopes that its genes may reinvigorate its cousin, the North Atlantic right whale. As Richard Sabin and field researchers such as Colin Speedie note, fin whales are seen in greater numbers in the Bay of Biscay and off the coast of southern Ireland, while blue whales are swimming through the Irish Sea, a passageway which once proved fatal for the easy access it allowed to British and Irish hunters, a kind of cetacean shooting alley. Taking advantage of the modern moratorium, the great whales are reclaiming their age-old routes.

However, this success makes them susceptible to those who consider their populations sustainable; ironically, our enlightened attitudes have exposed the whales anew. To take a thousand minkes has an exponential effect on the rest of the population by destroying complex breeding and social structures; the effect on sperm whales may be even more disproportionate. In 2006 Iceland announced its intention to resume hunting finbacks, although its efforts stalled with the discovery that the levels of mercury in the whales they caught were too high for human consumption; the Inuits of Greenland, who eat beluga and narwhal *muk tak*, are among the most contaminated people on earth, despite living in its least developed and apparently pristine spaces, while the whales in the Canadian St Lawrence waterway have absorbed so many industrial pollutants that one in four die of cancer. Norway, with its deep-rooted historical precedents, resumed commercial whaling in 1992. It never had any intention of abiding by the tenets of the IWC, nor does it see any contradiction in its actions: whales are as much livestock as any domesticated cow, a time-honoured resource for a maritime nation. Meanwhile, the contested moratorium remains in place, only ever a temporary solution, as both sides know only too well.

It took time for science to recover from Dr Lilly's extraordinary claims about cetacean intelligence; scientists were as loath to pronounce on the subject as they were to address the existence, or not, of the Loch Ness monster. None the less, it was becoming clear that whales and dolphins have brains matched only by the higher primates and humans, with whom they share the same convoluted neocortex – the characteristic wrinkles and whorls on the top layer of the organ – and which indicate exceptional intelligence. If allowances are made for their thick blubber, the body-to-brain-size ratio (the Encephalization Quotient, or EQ) of sperm whales indicates significant acumen.

Studies show that cetaceans can solve problems and use tools; exhibit joy and grief; and live in complex societies. Not only that, but they also pass on these abilities in 'cultural transmission'. Twentieth-century whaling may have destroyed 'not just numerous individuals', says Hal Whitehead, 'but also the cultural knowledge that they harboured relating to how to exploit certain habitats and areas'. The remaining animals also experienced lower birth rates as a result, and although they did not suffer as badly as mysticetes such as the right whale – which were reduced to a mere fraction of their pre-whaling numbers – the slow-breeding sperm whale population is growing at a mere one per cent a year. The 1986 moratorium may have come only just in time for *Physeter*.

Dr Whitehead – along with scientists such as Jonathan Gordon and Natalie Jacquet – has spent years studying the sperm whale in the wild. There is strong evidence that these whales are 'cognitively advanced', he tells me; they just don't use their brains in the same way as humans do. Their lives, lived in another medium and reliant on entirely different structures and influences to ours, demand other talents which are quite unknown to us.

Hal Whitehead's conclusions about the sperm whale are fascinating. He notes that while its brain is huge, it is not so unusual when seen in relative size, compared to other mammals. However, its structure 'suggests strengths in acoustic processing and intelligence'; it has an unusually large telencephalon, the area of the brain used to produce conscious mental and sensory processes, intelligence and personality, and its neocortex – associated with social intelligence in primates – is also highly developed.

It is precisely because the animal is so big, because its habitat is so huge, that its very existence provokes intelligence. Always moving, always in social groups, the whale's life is invariably interconnected, dependent on one another and on each other's knowledge. Its long, relatively safe life free from predators, and its great numbers have allowed the sperm whale to evolve elaborate social systems and cultures – although we are not quite sure what they are. And while Whitehead's research has found no direct evidence as to its intelligence – principally because so much of that life is unknown to us – the whale's complex social behaviour suggests a system of communal recollection, passing on information on feeding grounds and other memories. In an ever changing environment, there is an importance to the elderly, a kind of life insurance for the species.

It may be that whales remember more than we suspect; like the proverbial elephant, they may never forget. Research on humpback brains has also discovered the presence of spindle neurons, otherwise confined only to primates and dolphins. These cells – important in learning, memory and recognizing the world around and, perhaps, one's self – first appeared in man's ancestors fifteen million years ago. In cetaceans, they may have evolved thirty million years ago. This discovery places humpbacks with the odontocetes – sperm whales, orcas and dolphins – as sharing complex social skills of 'coalition-formation, co-operation, cultural transmission and tool usage'.

Hal Whitehead speaks of sperm whales as not only possessing a culture – the ability to learn information as a result of social interaction – but having used it 'to adapt successfully to the ocean's demanding environments'. 'There is a growing recognition that culture is not an exclusive property of humans.' Such research suggests entire communities of whales, ocean-wide clans moving in distinctive patterns and 'speaking' in distinctive repertoires of clicks, like humans sharing the same language. Separate groups of the same species will act in different ways, foraging for food in different manners – methods learned maternally, passed on from generation to generation. Similarly, Clan membership is comparable to nationality in humans; two clans off the Galápagos, although genetically similar and geographically close, 'talk' in different dialects.

Dr Whitehead organizes the sperm whale's clicks into four functional groupings: usual clicks, about two a second, made by foraging whales; creaks, a regular, more rapid succession of clicks which he describes as sounding like the rusty hinge on an opening door, and which indicate a whale homing in on its prey, or scanning other whales at the surface; the communicative sequence of codas – such as click-click-click-pause-click – a kind of cetacean Morse code which suggests 'conversations', although 'we do not know what information is being transmitted'. Most mysterious of all are the slow clicks or clangs made by mature males and which Whitehead compares to 'a jailhouse door being slammed every seven seconds'.

As scientists become aware of the complexity of sperm whale societies – which hunting may have fatally undermined by removing its matriarchs and, with them, essential knowledge needed to support the species, and by culling the large bull males with whom the females mate – those same societies face signal new threats. Even as we are able to predict the weather, it becomes unpredictable; as we find out more about whales, they start to disappear. Natural history may soon become simply that: history.

The earth's flora and fauna are vanishing at an average of one hundred a day. A process of extermination, first envisioned by Baron Cuvier two hundred years ago as man began to examine the nature of the whale, will be accomplished. Between beginning this book and finishing it, one species of cetacean, the Yangtze River dolphin, has been declared extinct. By the end of the century, half of all animal species – including the right whales of Cape Cod Bay – may follow the same route.

The sperm whale, too, has an uncertain future, so slow to breed that it also may eventually perish as a delayed effect of hunting. What man started, in the purposeful culls of the two great periods of whaling, his heirs may succeed in finishing off, almost by accident. Dr Whitehead and his fellow researchers may never know the truth about the whale; although in his most astonishing suggestion – all the more so for coming at the culmination of the most detailed scientific surveys, mathematical models and precise plottings, the result of a lifetime spent studying this one species – Whitehead proposes that we may yet discover that sperm whales, the oldest and possibly most evolved of any cetacean, have developed emotions, abstract concepts and, perhaps, even religion.

If sperm whales have religion, do they believe in us? In Melville's counter-bible, Ahab's blasphemous pursuit of Moby Dick ends in an apocalyptic, three-day chase. Driven to the edge of his mania, he plunges his harpoon into the animal's side – '*Thus,* I give up the spear!' – only for the rope to loop around his neck, 'and voicelessly as Turkish mutes bowstring their victim, he was shot out of the boat, ere the crew knew he was gone'. Ahab is last seen lashed to the whale's white side, as if crucified, his lifeless arm beckoning to the rest to follow him into watery oblivion. Then the animal turns on the *Pequod* and stoves in the ship, sinking her and all her crew. The entire human cargo of Melville's story disappears, leaving the surface as if man had never existed, 'and the great shroud of the sea rolled on as it rolled five thousand years ago'. Ishmael alone survives – clinging to a coffin made for Queequeg – to be picked up by a passing whale-ship, 'that in her retracing search after her missing children, only found another orphan'.

But what most commentators neglect to note is that there is another survivor from Melville's book: the whale itself. And if any animal were to evolve its own religion, what better animal than one that, for all its trials and tribulations, remains an immortal, omniscient power, a lingering shape in the ocean, beyond all human comprehension and physical dimension, forever spinning into space.

Timberline

Sloop

Roswell

Anchor

Whisk

Ventisca

Scratch's calf

Valley

XIII

The Whale Watch

> Could a greater miracle take place than for us to look through each other's eyes for an instant?
>
> Henry David Thoreau, *Walden*

At Macmillan Wharf, the boat is ready for her first whale watch of the day. Dennis Minsky, the naturalist, is looking over yesterday's survey, photocopied forms clipped to a board. He brushes his hand through his salt-and-pepper hair and smooths his moustache. Today he will compile another sheaf of data to be duly processed, pieces in a jigsaw that will never be completed.

Captain Mark Delumba stubs out his cigarette in a used coffee cup, then sets a course that is always the same but ever changing. As we lose sight of the Pilgrim Monument to thick mist, the sun disappears; all sound is muffled as the land falls away to the sound of a foghorn. We are the pioneers of the day; in our watery tracks the other boats will follow, bearing a mixture of children and parents, lovers and loners, the lost and the found, all looking for something.

It is a familiar sequence: the guano-spotted breakwater surmounted by heraldic cormorants and a lounging harbour seal, followed by a procession of lighthouses which mark our leaving of land's end: Long Point, where the sandy spit drops abruptly to one hundred and forty feet; Wood End, where a satellite of Provincetown once stood; and Race Point, where the water turns rough close to the deceptive green shallows of the shore. Vessels are often turned back here; to make it this far is an achievement. In the bay, the sea may be merely ruffled by the wind. Beyond the point, it can throw our hundred-foot boat about like a baby's bath-time toy.

The wind picks up as we pass into the open ocean. The depth gauge falls to eighty, seventy, sixty feet, indicating the rising presence of Stellwagen Bank below us, its shape a submerged echo of the Cape. This underwater plateau, an Atlantic Serengeti, is the epicentre of the food cycle that summons the whales, animals as migratory as any bird in your garden, Dennis tells the passengers.

It is a striking comparison: the airy bones of a flock of swallows, and the oil-rich bodies of a school of whales. Both travel equally vast distances, and this summer the returning whales are doing well. Sixty-eight cow-calf pairs have been identified, adding to some two thousand known individual humpbacks, testament to the wealth of these waters. None the less, they remain endangered: these animals' own ancestors were harpooned by whaling ships in the last century, and some may yet become targets themselves.

As the sea-bed is warmed by the sun, sand eels or lances wriggle out of their burrows towards the light. On the upper deck, Dennis shows his audience a rubber version of the same fish while his assistant holds up a series of instructive but rather heavy boards showing them what they might expect to see. The children squeal as Dennis passes the fish around, followed by a tiny specimen jar of sea water containing countless minute copepods. His final exhibit is an ancient clutch of baleen plates, fringed and brown and brittle, the colour and texture of horses' hooves.

Five years after I first boarded this boat, I have a different role. Now I'm part of the whale watch, rather than a mere observer. Through my field glasses, I loosen my eyes, letting them ride over the horizon. It is ever a nervous quest. I look for anything to indicate the presence of whales: subtle, hypnotic changes in the sea's surface which mark the meeting of currents; a flurry of gulls in search of a free meal; any little anomaly to break the monotonous view.

The boat ploughs ahead. Greater shearwaters, no bird ever so aptly named, tip their wings until they almost touch the waves, playing daredevil with the ocean; like Wilson's storm petrels, they are pelagic birds, spending all their lives at sea. Dennis quotes Aldo Leopold on how animals imply a landscape. Our captain swears loudly about the lack of fucking whales. Then, with eyesight sharper than my borrowed binoculars, he sees a distant blow. And with that, everything changes.

> To a landsman, no whale, nor any sign of a herring, would have been visible at that moment; nothing but a troubled bit of greenish white water, and thin scattered puffs of vapour hovering over it, and suffusingly blowing off to leeward, like the confused scud from white rolling billows.
>
> The First Lowering, *Moby-Dick*

There are strict regulations governing the approach of whales. From two miles away, speed must be reduced to thirteen, then ten, then seven knots; six hundred feet from the animal is the statutory stand-by zone. Even aircraft must maintain a minimum altitude of one thousand feet; John Waters quips that the whales are more demanding than Hollywood stars in their requirements for respectful distance.

In the wheelhouse, activity accelerates as the boat shudders to a halt. We bundle up cameras and clipboards to climb the ladder to the roof, where the captain takes control. As we glide towards the blow, the passengers let out their own gasps of excitement. Cameras click in a digital fusillade, but their lenses reveal only second-hand images of what their owners' eyes witness: creatures out of all scale with our world; animals so strange that sometimes it seems as if we hadn't seen them at all.

BE (Boston Entry) Buoy, 42°.14.88 N, 70°.17.45 W

The arrival of the leviathan is all the more surprising for its unassuming manner. As it rises, rivulets run off its graphite-black back like threads of quicksilver; huge pectoral fins glow below the surface, turned luminous green by suspended plankton. Slyly, it even shapes the sea in which it swims. The whale's mountainous body creates its own valley as it bobs in the ocean; while the pull of its tail leaves a slick of flat water as still as a pond even in rough seas. This flukeprint, the spoor by which its route can be traced, was believed by whalers to be oil or 'glip' washed off the whale as it dived; they would not cross it for fear of gallying their prey. The Inuit, too, decline to break its spell; but they do so out of respect, seeing this *qaala* – 'the path of the unseen whale' – as the animal's mirror into our world, and our mirror into its own.

Finback blow

As massive as it may be, a whale may be identified by its misty blow: the tall columnar geyser of a finback; the brief, staccato snort of a minke; the bushy blow of a humpback, the steam-engine of the sea sometimes turned into an indignant-sounding elephant trumpet; and the distinctive v-shaped spout of a right whale – for many whale watchers, the nearest they will ever get to seeing such rare animals. These are iridescent, airy signifiers of something so huge. They are also – as I discover only after I have been serially sprayed – capable of conducting 'flu-like infection.

Humpback blow

Little wonder that Tom, our videographer, turns his face away as the whale exhales again.

Other, more physical clues to the species are as subtle. The finback, for instance, is the only cetacean – indeed, the only mammal – with asymmetrical markings: one half dove grey, the other albatross white, an elegant division that extends even to its baleen, and which seems somehow to camouflage it in the ever changing light and shade of the sea. Its flukes, angular and sharply defined, are seen only when the animal lunge-feeds, moving on its side through a food source as it uses the white of its jaw to flash the fish into submission. Although their muscular backs are emblazoned with subtle swirls and chevrons, individual fin whales are usually recognized only from ship-strike scars; they are known by humans only by what humans have done to them. One animal, Braid, has what looks like tractor tracks across its back, result of an encounter with a propeller; perhaps Branded would be a better name. Another, Loon, has a white

scar reminiscent of a bird with a fish in its beak. As they surge through the water, their sheer length takes your breath away; a magnificence measured by the amount of time it takes for their never-ending bodies to slice through the surface. They are truly the athletes of the sea.

Humpbacks present a readier target, not least because they spend more time at the surface than almost any other whale. The underside of their flukes bears black and white patterns – partly a birthright, but often with notches and scars acquired during their lives at sea. By these marks, as distinctive as any human fingerprint, individuals are recognized on their return. New identifications are made as females bring their calves back to learn how to feed; only two years after it is first seen – time enough to survive long migrations, disease, or attacks by orcas – is a young whale given a name, inspired by the complex riffs and streaks on its flukes into which shapes are read like faces in flames or countries in clouds; a game which would no doubt appeal to the Prince of Denmark.

Added to these are other observations, such as dorsal fins spattered with white or extravagantly falcate – that is, sickle-shaped or tall enough to tremble as their owner moves through the water. Sexing the whale is quite another matter. The most obvious clue is the presence of a calf. Otherwise, only when glimpsed in a breach, or lying on its back like a sea lion lazily slapping its flippers from side to side, will a female reveal the gibbous swelling at its genitals; a male merely boasts a slit in which it stows its penis for hydrodynamics' sake. All these signs make a composite picture, assembled from snatched glimpses seen in or through the water as though through smeary glass, never quite complete.

Down below on the lower deck, passengers thrill to every fluke. Up above the wheelhouse, furious activity is under way. The crucial moment in a naturalist's encounter with a whale is, paradoxically, its departure. Abruptly and without warning, the humpback flexes its back and, deploying its massive muscles to lever its weight downwards, dives below. The movement is fluid, sinuous, of a piece: the rising and falling rostrum; the arching back and dorsal fin; the curving, sinewy tail and broad flukes, water dripping from their trailing edge, a diamond curtain glittering in the sunlight. The whale is freeze-framed in the act, caught at this tipping point between its world and ours.

At that instant of leave-taking, the animal presents its graphic ID, the markings on the underside of its flukes. If the view proves elusive, there follows a debate that can last for hours, perhaps even days. The captain is often the first to call a whale. After two decades at sea, Mark Delumba prides himself – despite his phlegmatic manner – on his ability to identify individuals, even at a distance. The Center's naturalist may be more circumspect. She or he will consult the onboard catalogue, a three-inch-thick file

illustrating the flukes of every humpback known here, arranged from mostly white to largely black in an abstract index of archipelagos and deltas and scars.

Back in the wheelhouse, the plastic-coated pages are scanned like a cop running a check on a young offender. The arguments continue until someone triumphantly stabs a fluke shot and calls out any one of a thousand exotic names: Ganesh, with a white patch resembling the elephant-headed Hindu god; Cygnus – a floppy, lopsided dorsal; or Colt, with her own pronounced fin and a habit of 'mugging' boats, remaining alongside for so long that captains call for other boats to lure the whale away so they can take their passengers home. Coral has the regular marks of orca teeth on her flukes, and a regular predilection for breaching and lob-tailing; Agassi has white spots on her fin, while Glostick boasts one white line on a mostly black fluke. Anchor has an eponymous anchor mark on the right-hand fluke; Midnight, a self-descriptively dark tail. Some whales, such as Stubb, Valley and Fulcrum, barely have dorsal fins at all, sliced off in inadvertent encounters with ships;

Nile, survivor of a brush with fishing line, has the white marks around her tail stock to prove it, while Meteor's right-hand fluke is ripped like a piece of paper torn to mark a page. It is salutary to see, after my years of watching these animals, how many of them bear such scars.

Most famous of all is Salt, the first of Cape Cod's whales to be named by Al Avellar, founder of whale watching in Provincetown, and known by her dorsal which looks as though it was sprinkled with the condiment. She is still bearing calves – humpbacks remain fertile all their lives – and is now a great-great-grandmother, with a family tree to rival any out of the *Almanac de Gotha.* Anchor's calf twists like an acrobat in the air, riding the wind as it rises from the sea, as if using its white pleated belly as a sail. Another, as yet unidentified, whale slaps its flippers so hard that it draws blood on the white skin where it has dislodged irritating barnacles. I can see the pink creases under its armpits, as if the animal were glowing from within. Such gestures seem almost lazy, but some researchers believe them to be acts of aggression.

Others clearly are not. Ventisca, notorious for her close boat approaches, rolls over and over on her back, flippers splashing me in the face; every ventral pleat is evident, running right down to her umbilical. When Nile sounds close to the boat, her sublime undercurve reveals a large melon-like lump; and as a young male swims by belly up, its genital slit is clearly displayed.

It is almost indecent to see the whales in such detail. I wonder, guiltily, if I ought to avert my eyes for fear of finding them too sensual – just as I admit that sometimes whales have the capacity to revolt me with their serpentine animalness, and I question why I invest so much of myself in them. Even now, I cannot reconcile myself to their corporeality. Then there are days when they seem all the more pristine for their appearance out of water, as if newborn. One afternoon, an unnamed yearling presented pure white patches on its upper flukes, its chin and over its eye, so sharply contrasted, one eyelid black, the other white.

For a few seconds – it cannot have been longer, although it seemed so – its eye met mine. Far from the dumb insolence of a horse or a dog's pleading fidelity, it fixed me with a stare which I found, and still find, disconcerting.

As three humpbacks travel across our bow, one is identified as Sockeye; it has a pronounced overbite resembling a salmon's. As it passes again, I notice a rope trailing from the animal's mouth, snagged in the baleen like a piece of floss; and as it sounds, I see that the line extends all the way back to its tail. It is entangled, a dog caught in its own lead without the wherewithal to free itself.

The passengers applaud the crew for choreographing them such a close encounter, but aloft the atmosphere changes. Karen Rankin, the naturalist, has already called the Center's disentanglement team, and as we return to Provincetown, the *Ibis* speeds out of the harbour, radioing us for details. I tell Scott Landry what I saw. Out at sea, the three whales surface close enough to allow the team to attach a grapple and slow Sockeye's progress, before cutting away ninety metres of gill net. In the process, the whale loses a tubercle – one of the sensitive, hairy nodes on its head – but it is a small price to pay for its liberty.

These are the new dangers the whales must face. At the peak of summer, there are near-misses with leisure craft. At one point, three finbacks are forced to dive abruptly under a cruiser that has drifted across their path. Karen issues a reprimand over our sound system; the offenders are lucky that Captain Joe Bones's decidedly less polite comments are not similarly broadcast.

A SELECT WHALE-WATCHING CAPTAIN'S GLOSSARY

Cameras!: Captain's exclamation on encountering good whales
Dutch boys: unexciting whales (as in watching paint dry)
Finback Alley: a stretch of water from Race Point to Peaked Hill, often frequented by *Balænoptera physalus*
Flashing: sight of the belly of a breaching whale
Hail Mary: a breaching whale
Lag: Atlantic white-sided dolphin (as in *Lagenorhyncus acutus*)
Mosquito: annoying civilian craft which tail the whale-watch boats
Mugger: close approaching whale
Old Bag: Salt, grande dame of these waters
Old Reliable: Loon, distinctively marked & frequently seen 'finner'
Pick my pocket: a whale stolen by another boat
Plastic: small boats, irritants (see also *Mosquito*)
Poison breach: a whale that breaches only once
Skunked: the condition of seeing no whales

Provincetowners all have their whale tales. Mary Martin, staying in an isolated dune shack, swam off Race Point one afternoon to find herself joined by a finback a hundred yards away. Jody Melander, driving on the winter beach in her truck, often sees right whales so close to the shore that she could easily join them, too. Some years ago a stray and possibly disorientated beluga appeared in the harbour, nudging curiously and dangerously around the boats' propellers. And in the summer of '82 a fifteen-foot female orca took up residency in the bay, its tameness indicating previous contact with man; some thought it an escaped navy trainee, used for military purposes. Pat de Groot kayaked out to sketch the animal, bobbing alongside it in her slender canoe, feeding it flounder – the fact that it took dead fish was a sign of its habituation towards humans – unafraid of its neat, regular, deadly teeth. Back in her beachside studio, she painted it again and again, in ink on flat grey stones. The whale stayed around, till one day someone decided that it needed a drink and poured whisky in its blowhole. It wasn't seen again.

The sea is the colour of steel and the sky. Peering over the prow, I see sand lances erupting at the surface, their dancing, silvery bodies punctuating the water like localized rain. Below them swims a school of bluefin, turquoise torpedoes darting this way and that, mouths open, voraciously.

Suddenly, a humpback swims under them all. Seen against its pale belly, the tuna seem minnows in comparison. The sand lances scatter like gnats. It is an animated lesson in the food chain. The whales distend their accordion pleats to swallow a ton of fish each day, even as daredevil gulls dive into their open mouths, or stride cockily up and down the whale's rostrum as if they were perched on a barnacled rock.

The ocean seems alive in these last days of summer. Three basking sharks swim by in silent convoy, their broad dorsal fins and razor tails swishing side to side, so different from cetaceans (it is part of the whale's essential *unfishiness* that its tail moves in the mammalian tradition). The sharks gape blindly as they feed on unseen plankton, their bodies browny yellow and mottled, almost reptilian, a mark of their own ancientness. A mola mola drifts along, a rudderless pancake of a fish, carried by the currents, warming its great flat body at the surface, its mouth opening and shutting as it gathers its food. White-sided dolphin, fleet and sleek, weave through the waves, a collective intelligence hoovering up the bait. Leaping high out of the water like competitive hurdlers, their teal and beige markings glisten in the sun.

The scene turns into a feeding frenzy, a sustained symphony of ecstasy. Humpbacks take great mouthfuls of fish which ripple across the surface in a futile attempt to escape. Finbacks lunge on their sides, flashing their prey with their white jaws. Minkes scissor through the same source, outriders to the greater whales. All around me is action, hunger, life and death, the entire natural cycle accelerated in a headlong dash for survival and sustenance.

The humpbacks gather below, blowing rings of bubbles in fine calibrations, their spiralling ascent announced by a bracelet of green clouds bursting at the surface. It is an unbelievably exciting moment, precisely because I know what is about to happen, announced by the changing colour of the sea, by the boiling cauldron of fish, and by the rousing *whoosh!* as the whales break through, mouths gaping like giant crows, close enough to see their bristly baleen and smell their fishy breath.

One afternoon I saw sixty or seventy of these animals gathered in a three-mile circle around us, a forest of blows and bubble-nets, some five or six animals in each group, each group multiplied by ten, and each surrounded by its own cloud of squawking gulls. Some were kick-feeding, a technique unique to Gulf of Maine whales, flexing and bucking their tails at the surface, smacking the fish into submission. The ocean itself seemed to be exploding. Our puny craft was completely diminished by the spectacle, a performance played out to a soundtrack of itself: a symphony of cetaceans, rising and falling with their own rhythms and unconscious beauty, repeated over and again in arpeggios of flukes and sinews and swollen throats so close together that they might push each other out of the water in their frantic efforts to feed.

Belly to belly, head to head, they were gathering up the bait as if they were afraid the food might run out. I could even see the sand lances leaping out of their mouths, in a futile bid for freedom. Delumba's father, who has fished these waters for forty years and has the missing digits to prove it, had never seen so many whales, too many to count or record. All we could do was to stand and stare as flukes broke the surface in every direction, leaving no square metre of ocean untenanted by a whale, each animal moving independently and yet in unison.

What struck me then, and does so even more as I try to reimagine what I saw, was the *surroundness* of it all; the fact that we were incidental to this act of ancient choreography; not so much spectators as prisoners, unable to move as we were encircled by the whales and their proprietorial blows. It was as if humans had never happened, as if the ocean had reverted to another Eden. We had to wait while they got on with the business of eating, of laying comprehensive claim to the world on which we merely floated. In their rising breath and dying fall all the power and poignancy of life seemed wrapped, fraught with dramatic suspension; an exchange of exhalation and inhalation which scares me to think of it. Yet that sound, which I can replay in my head even as I write, is also oddly consoling, a reminder of our common ground, a reassurance that everything will be all right, even if it will not. Perhaps whales will teach me how to live, just as my mother taught me how to die.

The same symmetry that had drawn me out of the city and back to where I was born had drawn the circle to a close: from the then when I needed my mother, to the now when she needed me; although she would never admit that she did, not in public, anyhow. She was fiercely independent, and would never submit. But I heard her on the phone to my sister, bemoaning her situation, and as creeping arthritis, which no amount of whale medicine would mend, added to her long list of ailments and her retreating senses, its crippling lock took hold of her legs and her fingers and her spine – even as I felt it in my own fingers. I overheard her, lying alone in bed, telling herself she would never walk again. She had always told me that at the end, when she was no longer needed, she would go down to Weston Shore, and just keep on walking. Now she couldn't even do that.

That September, soon after I had returned from the Cape, I was summoned by an early morning phone call to the hospital. My mother had suffered a severe heart attack. For a week, she lay there, slowly ebbing away on a hospital bed, with her family around her. At one point, I followed her as she was wheeled into intensive care, an air-locked, semi-darkened chamber where the blips and beeps of the other souls caught in limbo lay between life and death, emitting their own forlorn sonar. Only weeks before, I was a patient here, albeit for only an hour, my body sent through the claustrophobic scanner which knocked loudly like a poltergeist as it analysed my brain, trying to find the source of the eternal ringing in my ears, as if I were listening to some distant machinery. Now, in the same building, my mother lay wired up to her own machine, spread-eagled like an animal in an experiment, her long grey hair pulled tight by an elastic band. Her eyes never opened or closed, but she called my name.

In those days, the details of which only now seem to come back to me, I lived in the hospital, wandered its corridors, sometimes walking in the cemetery set, with shocking efficiency, across the road, where the early autumn sun shone low through the trees, their fading leaves filtering the light. Then, in the dark hours before dawn, as I awoke suddenly on the camp bed made up by her side, I heard her breath slow to an imperceptible halt, from being to not being, leaving me, another orphan. And as I bent over the bed – so quiet as if not to wake her – her mouth let out a final little gasp, just as mine gave its first fifty years ago.

> For now he was awake and knew
> No one is ever spared except in dreams
>
> W. H. Auden, 'Herman Melville'

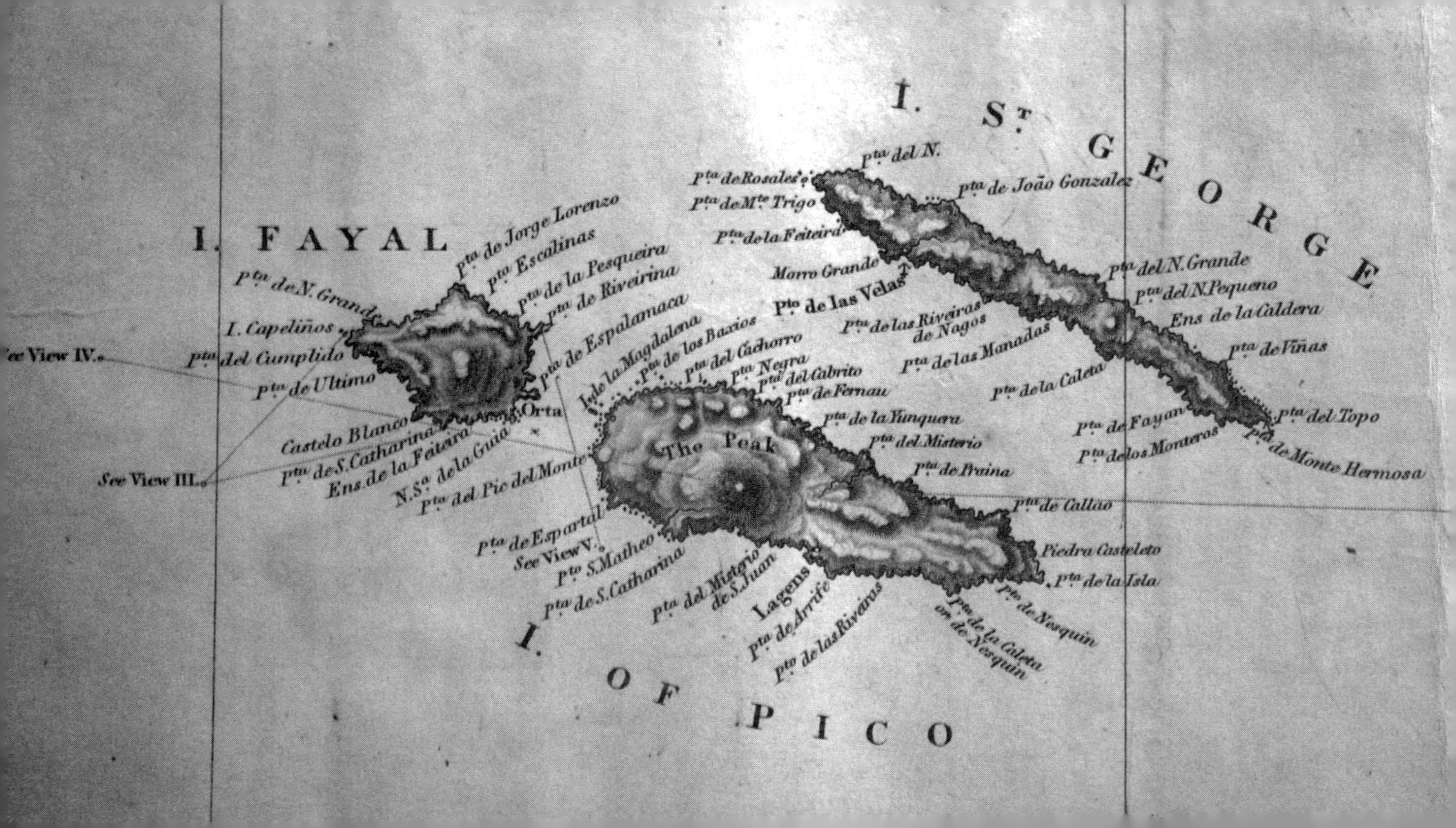
I. Fayal
I. St. George
I. of Pico
Pta de N. Grand
I. Capeliños
Pta del Cumplido
Pta de Ultimo
See View IV.
See View III.
Castelo Blanco
Pta de S. Catharina
Ens de la Feitara
N. Sa de la Guia
Orta
Pta del Pic del Monte
Pta de Jorge Lorenzo
Pta Escalinas
Pta de la Pesqueira
Pta de Riveirina
Pta de Espalamaca
Isle la Magdalena
Pta de los Baxios
Pta del Cachorro
Pta Negra
Pta del Cabrito
Pta de Fernau
Pta de la Yunquera
Pta del Misterio
Pta de Praina
Pta de Callao
Piedra Casteleto
Pta de la Isla
Pto de Nesquin
Pta de la Caleta or de Nesquin
Pto de las Riviras
Pta de Arrife
Lagens
Pta del Misterio de S. Juan
Pta de S. Catharina
Pto S. Matheo
See View V.
Pta de Espartal
The Peak
Pta de Rosales
Pta de Mte Trigo
Pta de la Feiteira
Morro Grande
Pto de las Velas
Pta de las Riveiras de Nagos
Pta de las Manadas
Pta de la Caleta
Pta de Fayan
Pta de los Monteros
Pta del N.
Pta de João Gonzalez
Pta del N. Grande
Pta del N. Pequeno
Ens de la Caldera
Pta de Viñas
Pta del Topo
Pta de Monte Hermosa

XIV

The Ends of the Earth

> The inhabitants are mainly of Portuguese descent, indolent and devoid of enterprise. Principal exports: wine and brandy, oranges, maize, beans, pineapples, cattle. The climate is recommended as suitable for consumptive patients.
>
> *The British Encyclopædia,* 1933

Fifteen hundred miles due east of Cape Cod and a thousand miles from Lisbon, the Azores lie in the middle of the Atlantic, scattered arbitrarily in the ocean. Portugal claimed these islands in the fifteenth century; Columbus called here to hear Mass on his way home from America. Most people would be hard pressed to find them on a map, falling as they do between the gutter of an atlas's pages. Yet these nine dots represent vast sea mounts greater than the Himalayas, a spine running the length of the earth in an invisible geography.

There are no friendly beaches of golden sand, only black rocks of bubbling lava arrested by the ocean. This is where the world is coming apart. Three islands lie on the Eurasian plate, three on the African, and the rest on the American plate; an act of perpetual tectonic division in which the westernmost isles inch closer to America and further from Europe each year. The youngest island, Pico, appeared only a quarter of a million years ago; its volcano is still active, and earthquakes occur here with fatal regularity. Sharply triangular against the sky, for Melville's Pierre, mourning the loss of his mother, it was an immemorial sight:

> Pierre hath ringed himself in with the grief of Eternity. Pierre is a peak inflexible in the heart of Time, as the isle-peak, Piko, stands unassaultable in the midst of the waves.

There is something foreboding about its outline, as though the entire archipelago were one enormous mirage. It was in Azorean waters that the *Mary Celeste*, Our Lady of the Heavens, was last seen in 1872, before being found abandoned with no trace of her captain or crew.

Each morning the ferry leaves from Faial, loaded with crates of supplies and passengers' luggage, borne across the narrow straits by waves that have travelled from the other side of the Atlantic.

They crash furiously over the rocks, rising in four-storey spouts and creating clouds of their own. But it is not the ocean's temper that fills me with trepidation; it is the fact that within a hundred yards it drops to a depth of one mile, and then far deeper.

It is a fear I feel as I walk through the dark streets of Lajes, past plane trees so severely pollarded that they look as though they are growing the wrong way up, stuck stump-down with their roots in the air. In the half light before dawn, the volcano blots out the stars, and somewhere over my shoulder the surf tears at the shore. This biblical little town is the oldest on Pico, perched on the island's southernmost shore and governed by two irresistible forces: the roaring sea and the restless earth.

At one end of Lajes is the tiny chapel of São Pedro, founded in 1460 and built into and of the basalt; at the other is a monumental eighteenth-century Franciscan monastery, its angles black-edged in mourning. Lajes is buttressed by belief, constrained by it. Its inhabitants are stocky, dark-eyed, yet also strangely familiar: they are the same handsome faces and the same names I know from Provincetown: Costa, Motta, Silvera. Even the taxi driver speaks English with a New Bedford accent.

Here too, whales are never far away. You see them in mosaics on the pavement, on souvenirs in shop windows, on the wooden fascias of cafés; one bar even boasts the toothless lower jaw of a sperm whale, suspended over its brandy bottles. Under the twin towers of the Santissima Trinidade, where Sunday-best children recite their catechism as their black-clad grandmothers sing, a glass cabinet holds scrimshaw models of harpoons pointing towards a crucified Christ set beside a little votive whale; a bone plaque dedicates these relics to Our Lady of Lourdes, whose miraculous appearance in a French cave in 1858 coincided with the commencement of whaling in the Azores.

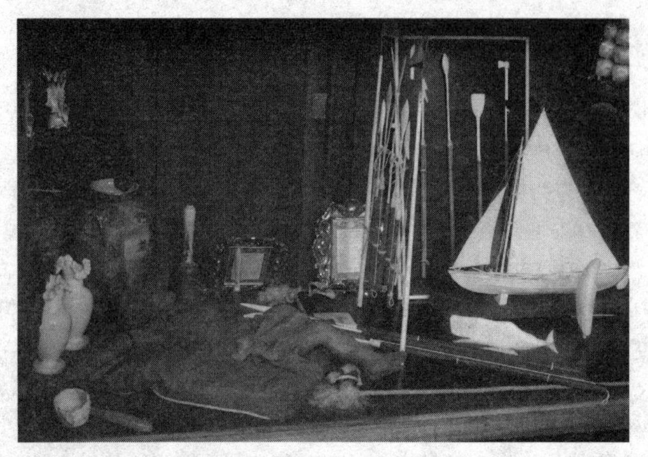

Homagere a Nossa Senhora de Lourdes dos baleiros dos Lajes do Pico.

If whales evolved long before humans then it seems fitting that they should still haunt these protean islands. The whales were here before the islands; and the islanders have lived off whales ever since the Americans came here in the mid-eighteenth century, sailing on the trade winds. Many ships – among them, the *Charles W. Morgan* – anchored in these waters, taking on fresh food and fresh crews. In turn, Azoreans worked their passage on a 'bridge of whale-ships' to the New World, as the same prevailing winds bypassed the islands on the voyage home, stranding Azoreans in America, where many made their homes; it has been calculated than half the population of the Massachusetts seaboard has Portuguese or Azorean blood. The islands themselves became archi-

tectural echoes of New Bedford and Nantucket, their narrow cobbled streets overlooked by rooftop lan-terns and clapboard; New England towns, only with palm trees.

Contrary to the claims of the *British Encyclopædia,* Azoreans are nothing if not resourceful, and in 1850 they began their own whal-ing. Soon one hundred Azorean crews were hunting whales, using techniques learned from their former masters. However, theirs is not a preserved memory of some distant past, for here, on these beautiful, diabolical islands, whaling did not end until 1986.

In a converted boathouse on the quayside, Serge Viallelle shows me film of Azorean whaling from the 1970s. It is like watching colour footage from the nineteenth century; as though Ishmael had a camcorder. The islanders used the same boats as the Yankee whalers, although latterly their double-prowed *canoas,* complete with whalebone cleats and trim, were taken out to sea by motor boats; and rather than spotting whales from the lofty crosstrees of a ship, they relied on *vigias,* towers perched on clifftop promontories where they still stand, just as wartime pillboxes still stud the southern coast of England.

Every morning the watcher would trudge up the narrow, flower-strewn path, his lunch packed in a neat wicker basket. Sitting on a wooden stool, peering through field glasses strapped to a swivelling stand, he would spend all day scanning the waves through the slit-like window, waiting for the blows that announced the whales.

At that sign the hunt began. The *vigia* would send up a rocket – lit by his cigarette – the signal for the crew to stop what they were doing. They might be digging in the fields or fishing at sea, but they were required by law to attend the call and liable to be fined if they did not. Like lifeboat men leaving their day jobs, they ran down to the harbour where their *canoas* stood ready. Once at sea, the men might spend all day and all night waiting for the whale. When it surfaced, they put up their sails and rowed silently towards the blow. This was the crucial moment. Unable to dive again until it had replenished the oxygen in its blood, the animal was at its most vulnerable in these, the last few minutes of its peaceable life. And all this was happening while I was going to nightclubs in London.

In the film, the irons find their target. The harpooned animal makes a forlorn dash, but, soon exhausted, it lies at the surface, where the lance is plunged again and again into its side; bent by the whale's struggles, the shaft is beaten straight on the *canoa's* boards before being used again. Blood swirls in the water, gouts of it; the whale shudders, and dies. Interviewed hunters testify to the excitement of the chase – 'Harpooning a whale is like scoring a goal' – a heroism worthy of the matador.

By the late 1970s each whale was worth £500; little wonder that subsistence farmers and fishermen were so eager to capture them. Yet whaling was truly a dying art. There was only one blacksmith left who could forge the harpoons and lances in their time-honoured shape. Even so, in 1979 one hundred and fifty sperm whales were caught off the Azores, and in the last ten years of whaling, the price of their teeth rose from three to eighty dollars a kilo.

Soon the islanders found better work elsewhere, and the world lost its taste for the products of the whale. The final blow came when the Azores joined the European Union, within which whaling was illegal. When Serge Viallelle came here from France in the 1980s, a drop-out delivering a yacht who discovered the whales and stayed, he had to persuade the islanders that people would pay just to look at whales. As in Provincetown, whale watching replaced whale hunting; in a neat twist of fate, the Azoreans were taught their new trade by Al Avellar, a Provincetowner of Portuguese descent.

In the nearby restaurant, the proprietor shows me through a mirrored door behind the bar, and into his sitting room. Its walls are lined with posters and photographs commemorating his years as a whaler. One shows him standing by a sperm whale, pointing to its huge teeth. He tells me that he killed twenty-two whales that year. As if to fill the silence as we stand in front of the picture, he says, 'People cry for the whales, but they do not cry for Iraq.'

For some reason, I pat him on the back. He says that whale flour was good for the crops, how they never had any insects when they were thus fertilized; no need for pesticides. Such useful things, whales.

Outside the restaurant, on the quayside overlooked by the volcano and the setting sun, an engine revs. Serge says it is the original motor boat that once towed the *canoas* out to sea. Whenever it starts up, he tells me, the sound scares the whales away for miles around.

On the north side of Pico lies São Roque. It has its own version of New Bedford's bronze harpooneer, posed with his weapon like an ancient Greek. Behind it a grey concrete ramp rises out of the sea, leading to a white-painted building with art deco lettering advertising its function:

VITAMINAS OLEOS FARINHAS ADUBOS ARMAÇÕES
BALEEIRAS REUNIDAS L.DA

It might as well be a factory on the outskirts of some Midlands town. But behind this façade lie blackened stone chimneys and abandoned outhouses; and in what appears to be an overgrown playground are the remains of a beached *canoa*, its splintered wood and fragments of whale bone held together by copper nails.

The main building is now a museum, although it is unlike any other I have seen. It is almost entirely empty: its exhibits are its fittings themselves. On the wooden walls are roughly chalked measurements and calculations. Under the high roof, vaulted with rusting girders, stand iron autoclaves as tall as a house. Buckets hang on hoists. The clang of metal doors all but echoes through this factory founded in 1942, as other factories were being built across Europe.

The men who operated these ovens left long ago. For half a century, sperm whales were taken from the seas around the island and towed here, sliding on their own blood and slime as they were winched up out of the water by machinery made in Tyneside.

At a cistern at the top of the ramp the head was drained of oil; the jaws were torn away and taken to one side. Then, in front of what looks like a garage forecourt with huge double doors ready to admit the beast, the rest of the whale was dissected.

Forty or fifty men in leather aprons and espadrilles went to work, slicing and sawing. Unlike their ancestors, they had the benefit of twentieth-century machinery. The blubber was wheeled in buckets to the ovens and rendered down in giant, hermetically

'Such is the endlessness, yea, the intolerableness of all earthly effort.'

sealed versions of try-pots. Spermaceti was kept cool in a concrete chamber, chilled by enormous refrigerated pipes.

In another part of the compound, whale meat was ground into flour for use as animal food. European cattle fed on whales. Nothing was wasted. This was the truly industrial, logical epitome of whaling. The whale's liver produced vitamin extracts. The teeth were used to create scrimshaw, destined to gather dust on tourists' shelves at home.

You could smell São Roque miles away, Serge's wife, Alexandra, remembers; it was a disgusting memory from her childhood. For the Englishman Malcolm Clarke, it was the stench of blood that hit you first. Then the sight of the severed jaws laid out to rot: 'The ground was literally alive with maggots.'

None of this is in the distant past. Men still bear the scars here, the teeth marks of whales on their bodies. Bones still lie on beaches.

A little way out of Lajes is a newly painted mural and a sign above what looks like the door of a garage: *Museu do Cachalotes e Lulas.* Inside is an eccentric collection, the product of one man's passion. Malcolm Clarke was born in Birmingham, grew up by the Thames, and spent his National Service in the Royal Army Medical Corps, driving ambulances from Aldershot to the military hospital at Netley on the shores of Southampton Water. In the 1950s he joined the whaling fleets of the South Atlantic and the Southern Ocean. His memory of that time is still vivid, and the numbers defy the imagination. In one season alone he saw thirty thousand whales taken. 'We were at full cook the whole time,' he says. Sometimes they caught twenty-four whales a day.

Malcolm became fascinated by what the whales ate. As we pass buckets filled with squid beaks, he tells me how the contents of sperm whales' stomachs would yield dozens of unidentified species; in one he found no fewer than 18,000 beaks. In fact, he now professes to find whales annoying, because they eat so many of the animals he studies.

The most impressive display in Malcolm's museum is a life-size cross-section of a female sperm whale painted directly onto the plaster, a mural so large that it carries on around the corner and onto the next wall. It is a lurid lesson in cetacean anatomy, but its bright blue and red organs cannot rival what lies on the table below. Swimming in a Tupperware dish is a sample of the spermaceti sac, glistening like tripe. I prod it, gingerly; the oil has crystallized like old honey.

Next to it is a square chunk of blubber. I am taken aback at how hard it feels, more like wood than fat. I squeeze a piece between my finger and thumb; the intricate mesh that runs through it barely yields. I imagine an armoured animal, tank-like.

'They were tremendously difficult for the whalers to cut,' says Malcolm. The blubber is also burrowed and wormed by parasites, a certain source of irritation for their unwilling host.

Something stranger lies in a third container: what looks like a lump of brownish-grey mud at the bottom of an old coffee jar. As I lift the lid, the smell hits my nostrils: pungent, musky, discernibly animal, its congealed, peaty texture reminds me of nothing so much as cannabis resin. Then Malcolm shows me on his diagram where this stuff came from: the whale's rectum. I am holding a piece of ambergris the size of a small potato, the most precious product of any animal, a natural creation more elusive than any gold or diamond. But what I had hitherto assumed to be the result of some mysterious process, like grit in an oyster shell forming a pearl, is actually whale shit.

It was a marvellous irony, thought Thomas Beale, 'that a resemblance to the smell of this drug, which is the most agreeable of all the perfumes, should be produced by a preparation of one of the most odious of all substances'. On his own researches into the interior of the whale, Beale cited the chemist Wilhelm Homberg, who found 'that a vessel in which he had made a long digestion of human fæces, acquired a very strong and perfect smell of ambergris'. This somewhat unsavoury experiment – which swiftly led to the evacuation of Homberg's laboratory by his assistants – brought Beale to the same conclusion: that ambergris was 'nothing but the hardened fæces of the spermaceti whale, which is pretty well proved from its being mixed so intimately with the refuse of its food'. Indeed, his friend Samuel Enderby possessed 'a fine specimen . . . about six or seven inches long, and which bears very evident marks of having been moulded by the lower portion of the rectum of the whale'. And during his own adventures in the North Pacific, Beale himself had collected some 'semi-fluid fæces' which had floated from the carcase of a whale, 'and which on being dried in the sun bore all the properties of ambergris'.

The exact origins of ambergris remain obscure; but it is certainly the result of a remarkable process. The sperm whale swallows squid alive, taking its food into the first of four stomachs. It then passes into a second stomach to be broken down by strong acids, assisted by a writhing mass of nematode worms, 'a disgusting sight' according to Malcolm, who has seen it many times. When the waste moves through the lower intestine the brittle, shiny black squid beaks – along with other indigestible material such as nematode cuticles – prompt the whale's digestive system to secrete bile and thereby ease their passage. Occasionally – in as few as one in a hundred whales – this chemical reaction produces ambergris. Once expelled, it may spend months or even years in the water, oxidizing and hardening into layered lumps, often still containing bits of squid beaks. Lighter than water, ambergris is occasionally cast up on beaches – hence its name, grey amber, an allusion to the fossilized tree resin also found on seashores.

Early authorities thought that ambergris was only produced by ailing whales. Frederick Bennett concluded that animals that displayed 'a torpid and sickly appearance' and which failed to 'void liquid excrement' when alarmed or harpooned were those most likely to yield the stuff. He reasoned that the sharp beaks could cause a cicatrix to form, a scarred wound which closed up the return, leaving the whale to waste away to its death, 'a goose killed by the golden egg within'. Modern cetologists, however, think that ambergris comes from healthy whales.

I smell the lump again, trying to detect its complexity like a wine-taster – the qualities that make it so desirable to *parfumiers*:

its ability to absorb, intensify and capture volatile fragrances, sometimes for years. It is as though its depth can encompass all aromas. As I hold it in my fingers, Malcolm warns that it will stay with me for days. I smear a little in my journal; months later, it is still there: the lingering scent of whale.

This romantic stuff – which reminded one scientist 'of a cool English wood in spring, and the scent you smell when you tear up the moss to uncover the dark soil underneath' – had many strange and exotic uses. The ancient Chinese called it *lung sien hiang*, or 'dragon's spittle fragrance', and spiced their wine with it. During the Black Death, ambergris was carried to ward off the plague. In the Renaissance it was moulded, dried, decorated and used as jewellery; it was also said to be efficacious as an aphrodisiac, as a medicine for the heart or brain, and for diseases such as epilepsy, typhoid and asthma. In Milton's *Paradise Regained*, Satan tempts Christ with 'Grisamber steamed'; and drawing on Thomas Beale's researches, Ishmael notes that the Turks took it to Mecca, 'for the same purpose that frankincense is carried to St Peter's in Rome'. More prosaically, sailors used it as a laxative.

Although Ishmael declares that it was whale oil that was rubbed on the British sovereign's head in the coronation service, this was in fact an ambergris-infused concoction, as I discovered on a visit to the Gormenghast-like library set in the eaves high above Westminster Abbey. Here the custodian of the panelled eyrie, reached by a door set in the gloomy corner of the cloister and at the top of a flight of narrow wooden spiral stairs, divulged to me the secret recipe, handed down over the centuries. '*Oleaum Præscriptum Ad Ungendum in Coronatione Carolum I Britanniæ Regem*'. Among oils of jasmine, rose, cinnamon, musk and civet was the all-important and precious ingredient, '*Ambrægrisiæ* ℥iiij', which created a fluid with 'a rich and peculiar fragrance; it is amber coloured when freshly made, but time deepens the colour and the odour becomes mellow and rare'. In the most sacred part of the ceremony, shielded from the common gaze by a canopy of cloth of gold, the new monarch is marked on the head, heart, shoulders, hands and elbows with this oil, although Queen Victoria is said to have hated the stickiness and the smell and insisted on washing it off soon after, rather than allowing it to baste her imperial majesty with its whale-stink.

This amazing substance remained as rare and mysterious as the unicorn's horn until the American whalers began to find it within the whale itself. In 1724, Beale records, Dr Boylston of Boston wrote to the Royal Society in London, having interviewed Nantucket whalers who 'cutting up a spermaceti bull-whale . . . found accidentally in him about twenty pounds' weight, more or less, of that drug; after which, they and other such fishermen became very curious in searching all such whales they killed, and it has been since found in lesser quantities in several male whales of that kind, and in no other . . .'

‘They add further,’ Boylston noted, ‘that it is contained in a syst or bag . . . nowhere to be found but near the *genital parts* of the fish. The ambergris is when first taken out moist, and of an exceedingly strong and offensive smell.’ The idea that this sac was situated at the root of the whale’s penis, along with the masculine smell as it ripened, contributed to the erroneous and perhaps chauvinist notion that only bull sperm whales could produce ambergris. Although males, being larger, produced bigger pieces, females were equally able to excrete their own perfume.

In 1783 Joseph Banks presented a paper to the Royal Society by Franz Xavier Schwedier, a German doctor, which conclusively identified the true origins of ambergris. The subject was even discussed in Parliament; and in January 1791 *The Times* noted that ‘a whale lately brought from the South Seas, in the Lord Hawkesbury, contained near four hundred ounces of ambergrease, which sold by the hammer at Lloyd’s Coffee-house at nineteen shillings and sixpence the ounce’, a great price to pay for this prize.

Like a precious metal, ambergris has retained its worth over time. In 1912 a Norwegian company was saved from bankruptcy by a one-thousand-pound lump found in a whale caught off Australia, and which sold in London for £23,000. In 1931 – as a cutting stuck inside my edition of Frank Bullen’s *The Cruise of the Cachalot* notes – a seventy-foot male found dead on New Zealand’s South Island produced a quarter of a ton of ambergris worth over £10,000. In the 1950s four pounds of this ‘floating gold’ fetched £100,000. Meanwhile, Soviet fleets gathered so much ambergris – including sixty-three pieces found in one whale – that by 1963 the Communist state no longer had any need to import it.

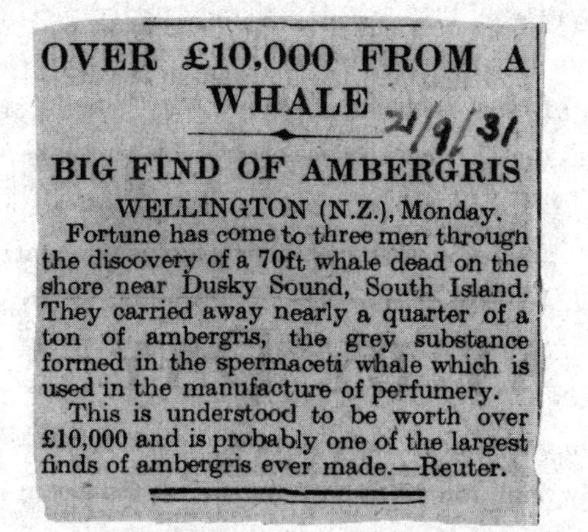

OVER £10,000 FROM A WHALE

21/9/31

BIG FIND OF AMBERGRIS

WELLINGTON (N.Z.), Monday.

Fortune has come to three men through the discovery of a 70ft whale dead on the shore near Dusky Sound, South Island. They carried away nearly a quarter of a ton of ambergris, the grey substance formed in the spermaceti whale which is used in the manufacture of perfumery.

This is understood to be worth over £10,000 and is probably one of the largest finds of ambergris ever made.—Reuter.

Modern chemical analysis would show that the active element of ambergris is ambrien, a crystalline, fatty cholesterol able to fix volatile oils by slowing evaporation. Despite synthetic substitutes, it remains an irreplaceable ingredient in perfume. All the grandest French houses still produce exquisite scents based on this most mysterious of components, from Chanel and Yves Saint Laurent to Givenchy and Christian Dior; if you happen to be wearing *Dioressence* today, you are wearing the scent of a sperm whale. One of the oldest perfumers, Creed's of London, which guards its formulæ as jealously as the custodians of the coronation rite, has been patronized by George III, by the Prince Imperial, dandy son of Louis Napoleon, who was wearing Creed's whale-infused fragrance when he met his death at the end of eighteen assegais in the Zulu Wars in 1879 – and by Cary Grant, for whom the company designed a perfume all of his own, based on ambergris.

Having smelt the raw material, I can now identify the trace of ambergris in expensive scents that waft from the shoulders of party-goers. Like their clients, perfume makers are, of course, discerning in what they buy. The highest prized pieces are pale in colour, from white to gold to grey with sometimes a mauvish tint; dark brown or black lumps are of a lesser worth. Most ambergris comes from the Indian Ocean, but when Dorothy Ferreira of Montauk, Long Island, inherited a large piece from an elderly friend, she was told that her gnarled legacy – which prompted a headline in the *New York Times*, 'PRECIOUS WHALE VOMIT, NOT JUST JUNK' – would fetch $18,000. And in a story which might have come from the pages of Roald Dahl, a ten-year-old girl found a yellowy lump of 'whale sick' on a Welsh beach, supposedly valued at £35,000. 'We recently heard on the radio about ambergris,' her mother told a tabloid newspaper, 'but when Melissa found some I couldn't believe it!' Unfortunately for Melissa, such finds usually turn out to be industrial plastic, or surfboard wax, or, as Richard Sabin reports, 'something even less pleasant'.

Yet even scientists have been known to become childlike when faced with the prospect of this elusive stuff. One told me how, when dissecting a sperm whale washed up on the island of Malta – a week-long process which began with some twenty-six cheerful helpers on the first day, but which had dwindled to a mere handful of hardy souls by the last, such was the stench – he squeezed through two hundred metres of malodorous guts in a determined but ultimately unsuccessful search for ambergris.

Light-giving wax, lubricating oil, scented fæces: sometimes it seems as though the whales are cetacean Magi, bringing offerings that presage their own sacrifice. Such is the whales' abiding paradox that they should secrete such precious substances from the profundities of their bodies, places as unknown as the seas in which they swim, even as our own interiors are a mystery to us.

Like Melville writing about Nantucket, an island he had never visited, I write about animals I have never seen, for all that I can smell them and handle their most intimate secrets. The closer I get, the further away they seem; and the more I learn, the less I know about these strange cetaceans, mammals like us, yet so separated in scale in our microcosms of greater unknowns, from the sea to infinity.

Even their most basic mechanics have a functional, fatal beauty. In Malcolm's museum, a diagram shows how a sperm whale's trachea and œsophagus share the same internal space, the one able to shut off the other to prevent the lungs filling with water as the whale feeds. Another charts the spectrum visible to deep-diving whales, which have blue-shifted eye pigments – being the most useful colour to discern in waters which turn from turquoise to black as they recede from the sun. A hinged wooden model demonstrates Malcolm's theory of how the sperm whale adjusts its buoyancy by altering the temperature of the oil in its head, although he allows rival scientific interpretations: pre-eminently, that the spermaceti functions as a focus for the whale's sonar clicks. A piece of bone cut in half shows honeycomb cells which in life would be filled with oil; filled with air, they would expand with changing water pressure as the animal dived.

So many occupational hazards for the whale.

There is something atavistic about these objects. The smallest comes from the whale's inner ear, the same shell-like bone found in the bilges of the *Morgan.* These are the parts of the whale that survive the longest: otoliths, the fossilized ears of fifteen-million-year-old whales, have been found in South Carolina, their strange curling chambers evocative of ancient oceans and prehistoric sounds, as if by holding them to your own ear you might hear extinct animals singing in long-vanished seas.

Outside his museum, on a rocky ledge overlooking the ocean, Malcolm has built a life-size model whale out of tubular grey scaffolding. It resembles a cross between Ishmael's Arsacidean temple and a children's climbing frame. As buzzards hover overhead, we talk about Malcolm's years at sea. At my urging, he even speaks of monsters: of the giant squid that one fisherman saw alongside his boat, its tentacles longer than the hundred-foot vessel, making the entire animal twice its length; and of the pilot of a whale-spotting plane flying over the Indian Ocean off Durban who saw a wrecked plane's fuselage sticking out of the water, only to watch the shape animate itself into a long neck and slip silently into the ocean.

Such stories seem to suit this infernal island, a half-formed place of fire and water; I could imagine Melville and Hawthorne meeting here. Even the cliffs on which we stand are undermined by hidden caves. Due south from here lies Antarctica. And somewhere down in the fathomless, gathering darkness, sperm whales swim, eternally aware, their lives one waking dream, moving through valleys that run thirty thousand miles along the ocean floor, through lakes that lie stilly in the abyss, separated by temperature like pools of mercury, past jellyfish pulsating as ghostly Victorian brides in ectoplasmic crinolines.

XV

The Chase

And I only am escaped alone to tell thee.

Job

'Now, Philip.'

João's command is urgent, unexpected. There is no time to get into my wet suit. I scramble to spit into my mask and jam the snorkel into my mouth. Marco stands on my fins so that I can push my feet into them. Climbing over the side of the rigid inflatable boat, I am launched into the Atlantic.

I am swimming in waters more than two miles deep. I can't see ahead of me. Below the blue gives way to complete black, the kind of impenetrable black I only ever saw in a cave in Cheddar Gorge as a child, when the guide turned the lights out and told us that we would never experience such a profound darkness.

João shouts directions from the boat. It is getting smaller with every minute that I swim away from it, away from safety, into the unknown. I might as well be swimming into outer space.

I hardly knew it, as we hurried to leave the harbour, but the conditions were perfect. The sea's surface was glassy, barely rippling in the summer sun. João, with his cropped hair and an orca tattooed on the calf of his leg, scanned the horizon through his sunglasses; Marco, his first mate, peering in the other direction as he hung from the superstructure of the rib, a modern-day whaleboat with a 250-horsepower engine.

As we picked up speed out of the harbour, a pod of common dolphins had zoomed out of nowhere and into our path, playing at the bow. Competing to be first, they rode so close I could easily have reached out and touched them. Steel blue and dove grey, their hour-glass, go-faster stripes were raked with each other's teeth marks; as cute as they looked, these animals were bigger than me. They swam in water so clear that they appeared to float in a vacuum, streams of silver bubbles trailing from their blow-holes. As they twisted and turned their bodies to peer up at us, it seemed as though they were escorting us to some appointed meeting, here at the end of the world.

Then something larger loomed ahead. Even from a mile away, I knew it was a whale – albeit one unlike any other I had ever seen before. Its blow was utterly distinctive, spouting at forty-five degrees to the surface. Instantly I saw the reason for the Latin name. It really was a big-headed blower; *Physeter* even sounded like its bursts of exhaled breath.

As the boat drew closer, I could make out a grey shape, lying like a pale shiny log in the water. It was difficult to tell one end from the other: which was the head and which the dorsal fin? Then, as it rose to take its breath, I saw its single nostril, wantonly lopsided. It was shockingly strange. The animal was an arrangement of sun-burnished bumps, 'compared to little else than a dark rock, or the bole of some giant tree', as Frederick Bennett wrote in the 1830s.

As the head lifted out of the water, I saw that it was not alone. Quietly, it became part of an assembly. Further off lay two or three animals, then yet more until, almost disguised by the waves, a group of ten or twelve sperm whales hung there, breathing in rhythm with the sea, a rhythm that caught my breath too as the boat rose and fell with the swell.

All that was five minutes ago, but it might have been a lifetime. Now I was fighting for air in the water, trying to remember to breathe through my nose and not my mouth, like them.

'To your left, Philip!' João called through cupped hands. I had no idea which was left and which was right. I kicked my legs furiously, but I didn't seem to be going anywhere. The waves

seemed to push me back and under. With my heart pounding in my ribs, I took a deep breath and peered beneath me, down into the unknown.

It was as if I were looking into the universe. The blue was intangible yet distinct; untouchable and all-enveloping, like the sky. I felt like an astronaut set adrift, the world falling away beneath me. Floating in and out of focus before my eyes were a myriad of miniature planets or asteroids, some elliptical, some perfect spheres. Set sharply against the blue, the glaucous, gelatinous micro-animals and what seemed to be fish roe moved in a firmament of their own, both within and beyond my perception.

I was moving through another dimension, suspended in salt water, held over an earth that had disappeared far below. I could see nothing ahead. The rich soup on which those same tiny organisms fed combined to defeat my sight, reducing lateral visibility as they drifted like dust motes caught in the sunlight.

Then, suddenly, there it was.

Ahead, taking shape out of the darkness, was an outline

familiar from words and pictures and books and films but which had never seemed real; an image I might have invented out of my childhood nightmares, a recollection of something impossible. Something so huge I could not see it, yet which now resolved itself into reality.

A sperm whale, hanging at the surface. I was less than thirty feet away before I saw it, before its blunt head, connected by muscular flanks to its infinite, slowly swaying flukes, filled my field of vision.

In a moment that seemed to go on for ever – catching my breath in my perspex mask, my limbs frozen with panic and excitement, my body held in suspense, not wanting to go forward but never wanting to go back – the distance between us closed.

Its great grey head turned towards me, looking like an upright block of granite, overwhelmingly monumental. Its entirety was my own. That was all I could see: far taller and wider than me, the front end of an animal which, it suddenly occurred to me, had one major disadvantage over the puny human swimming towards it. It could not see me. Its eyes could not take me in. I was approaching the whale from its blind spot. And it was coming closer.

What if it just kept on coming? The head bent down, bringing its ponderous dome to bear in my direction. Then I began to hear it.

Click-click-click, *click-click-click*, click-click-click.

A rapid series of sounds, creaking. I felt them, rather than heard them, in my breastbone; my ribcage had become a sound box. The whale was creating its own picture of me in its head; an MRI scan of the intruder, an outline of an alien in its world.

I felt my body let go, and peed into the water. A ridiculous thought passed through my mind: I had arrived unannounced, only to lose control of my bodily functions and piss on my host's doormat. Then, at the crucial moment, the head turned, bowing slightly, as if in identification. Not edible. Not interesting.

From sheer fear the moment turned into something else. I realized that this was a female. A great mother hanging before me, intensely alive. For all her disinterest, it seemed there was an invisible umbilical between us. Mammal to mammal; her huge greyness, my unmothered paleness. Lost and found. Another orphan.

I could not believe that something so big could be so silent. Surveyed by the electrical charge of her sixth sense, I felt insignificant, and yet not quite. Recreated in her own dimension, in the dimension of the sea, I was taken into her otherness, my image in her head. As the whale turned past me, I saw her eye, grey, veiled, sentient; set in her side, the centre of her consciousness. Behind it lay only muscle, moving without effort. The moment lasted for ever, and for seconds. Both of us in our naked entirety, nothing between us but illimitable ocean.

Then she was gone, plunging soundlessly into the black, silhouetted against the blue, her shape so graphic it might have been created by computer, a CGI image set against a cinematic matt. Only as the distance between us increased – as the silence of her descent became hypnotic – was her ancient enormity revealed; something I had seen, and yet which I could not quite comprehend.

Back at the boat, Marco hauled me out of the water, and João smiled, shaking my hand and saying solemnly: 'You are a lucky man.'

Over the next few days, I spent all my time at sea, beyond the land. I didn't need my credit card or my keys. While people were shopping, eating, talking, waking, sleeping, I swam with whales.

Often I could not see the whales as I entered the water, and had to trust entirely in João's shouted directions. Sometimes the animals would move so fast that they vanished before I could swim within sight. I watched their diminishing shapes, a trio of whales with their tails moving almost imperceptibly, powering them into the blue. But sometimes I found myself closing in, moving towards huge heads rising rhythmically with each blow as I blew air out of my own snorkel. I saw them on their level, rather than from above, as the great flukes rose on tails drawn vertically out of the waves – the hand of God so feared by the whalers – before plummeting with immense grandeur into the deep. I was within their world, rather than outside it; looking into it, rather than merely looking on.

Then the bad weather rolled in, and for days the seas lashed Pico, smashing white against the black rocky shores. The boats lay tied up in the harbour. At night, the Cory's shearwaters, which by day followed the whales like courtiers, came in to roost, their ghostly shapes circling over the darkened harbour, singing almost comically, 'sqwhack, sqwhack, sqwhaaackkk'.

I lay in bed, unable to close my eyes. Every time I did, I would see the whale. All my life I had dreamed about whales. Now the void had been filled; or rather, I had been taken into it. What was I trying to prove? All my fears of loss and abandonment and being left behind seemed to be summed up in this confrontation, so extreme that it induced a state of suspended hallucination. As I lay sleepless in a hired bed, I thought I might lose my senses entirely, in the early hours of the morning, the same time when I had lain on the floor in the hospital ward, listening to the breath that had brought me into the world slowing to a halt.

Then, in the seeping light of morning, with the volcano looming out of the dawn beyond my window, the sea abruptly calmed, like a hand stretched over its surface.

Lowering the hydrophone over the side, João listened intently to the clicks echoing over the ocean. Beneath us, under the thin floor of the boat, the whales announced their position, the clicks increasing in intensity, in patterns I could not discern –

click – click – click ~ click-click-click ~ *click-click-click*

– accelerating ever closer, entirely in charge of a world over which we floated. It was as if they were resounding miles below, even as they radioed their presence to other whales miles apart. Tuned in to some unseen circuit of food and communal intent, they knew instinctively where they were, while we wonder constantly what on earth we are doing.

A smooth rounded shape ploughed through the water towards us, its melon and pointed beak the unmistakable form of a beaked whale. 'For me, it was a Sowerby's,' said João. It was a species I knew only as a model in the museum or a picture in my handbook: '*Mesoplodon bidens.* Status: unknown; Population: unknown; Threats: unknown'.

The seas were alive off this island of rarities. Suddenly, animals were everywhere, conjured up out of the otherwise empty sea, as though one of my manuals had come to life. The sheer variety was astounding. Pods of oceanic striped and spotted dolphin raced past, their markings like fine china, followed by a school of short-finned pilot whales, calves swimming so close to their mothers' sides that they seemed attached by invisible strings. A manta ray swam under our keel, like a great bat. Marco picked up a passing hawks-bill turtle. It eyed us suspiciously before being released back to the sea, where it paddled incongruously like an overgrown tortoise.

The life leapt out of the ocean: as we sped past the vertical cliffs on which the *vigia* stood, something butterfly-like shot out of the waves, level with my eyes – a flying fish with rainbow wings, an unreal, glittering invention like some fantastical clockwork toy. Even the sea's surface was decorated with drift-ing Portuguese men o' war, their inflated bladders edged with a fluorescent pink frill, trailing colonies of magenta and purple

tentacles, each an animal in its own right. I wanted to reach out and right the aimless creatures as they were blown over by the wind like lost balloons, although I knew my reward would be a potentially lethal sting.

Ahead, there were blows. The whales had returned, drawing deep breaths on long dives in search of food. As they passed us, a red, ragged lump floated to the surface: a giant chunk of left-over squid, its tentacles torn like meat fed to lions in a game park.

One whale lunged on its side, close to starboard, its pale speckled jaw visible through the water. Another slowly raised its squared-off snout as it spy-hopped, bringing its eyes level to look at us, even as we looked at it; at that moment, the entire animal was hanging vertically in the ocean, perpendicular to the surface.

These details were lessons in the natural history of sperm whales; I was being given a personal crash-course in practical cetology. Often I saw the animals' wrinkled flanks, bark-like creases running from head to tail, bodies puckered as though they had spent too long in the water. Coming upon a group of three females, the adults dived in sequence, leaving their calf behind as if we were baby-sitters. When they resurfaced – a grey flotilla with heads rising as prows – to collect their charge, it seemed we had got too close, and the nearest adult slapped the surface sharply with her flukes, warning us to keep our distance.

These waters were their home: their nursery, their living space, their dining room. One whale raised its flukes and squirted out a cloud of reddish poo, rank with the odour and colour of digested squid. Another left behind a sliver of sloughed skin. João scooped it up out of the water and gave it to me. It had the same colour as the whale, but was gossamer-thin, lying like skein of grey snot in my hand. Later, I laid it on a page of my journal, where it dried to a tissue yet smelled as strong as ever – the 'peculiar and very strong odour' which impressed Beale, and which Ishmael could smell from miles away, 'that peculiar odor, sometimes to a great distance given forth by the living sperm whale, was palpable to all the watch'. It was also deeply male and musky, strangely sexual and arousing, like the little bottle of sperm oil I found on a shelf at Arrowhead.

> . . . you may scrape off with your hand an infinitely thin, transparent substance, somewhat resembling the thinnest shreds of isinglass, only it is almost as flexible and soft as satin; that is, previous to being dried, when it not only contracts and thickens, but becomes rather hard and brittle. I have several such bits, which I use for marks in my whale-books. It is transparent, as I said before; and being laid upon the printed page, I have sometimes pleased myself with fancying it exerted a magnifying influence. At any rate, it is pleasant to read about whales through their own spectacles.
>
> The Blanket, *Moby-Dick*

I never failed to thrill to the appearance of the whales in those days at sea. Falling in synch with their cycle, with the swell of the sea, I came to know when to expect their arrival, and when they were about to leave. Hour after hour we would wait for them to surface; sometimes I would lie in the prow of the boat, from

sheer exhaustion, falling asleep in the sun – only to be roused by the appearance of another animal: the plosive announcement as it arrived, its rounded head breaking the surface; the minutes it spent 'rafting', lying like a panting dog catching its breath after a run. Then the head would rise as it took its last breath, the body straightening briefly before the back arched, the great ridge of knuckles flexing beneath the taut skin like a resurgent mountain range. Finally, the animal pulled up its tail and levered itself into the ocean.

This announced sequence, invariable and majestic – the muscular tail, so much more upright than any other whale I had ever seen, like some vast grey tree trunk; the powerful backbones exposed, just as the colour of your bones is revealed when you clench your hand to make a fist; the trailing edge which announced the individual's identity – all this was constantly strange and exciting. It induced a state of perpetual nerves: to be witness to this repeated beauty was almost too much. Yet there was also something immemorial about the articulate right-angledness of its leaving, the flexibility of something so huge – the presentation of the distinctive shape of its flukes which marked it out, their geographic lines echoing the island on the distant horizon – before vanishing with barely a ripple, so sublime was the animal's re-entry. It was at this moment that the whales seemed at their most dinosaurian, most prehistoric; it was easy to believe, at such times, that these creatures were older than any other. Then the waiting began all over again.

Ah the world, oh the whale.

All day I sat in my wet suit, as rubbery as the side of the rib, nervous, ready. Two or three times there were false alarms as João could not get his boat ahead of the whales; to approach from any other angle would be futile, as his predecessors knew.

The sun beat down, turning my body brown, tattooing my neck and wrists with tidelines to remind me of my encounter. The waves lapped languorously at my feet as I dangled them over the side. I wanted to get back in.

'Let's go.'

This time I was ready, protected against the deceptive chill of the sea; insulated, like the whale. I dropped over the side, fingertips leaving go, letting my body bob in the water and find its own buoyancy. João's shouted directions drifted away with the boat. I was left alone, moving steadily towards the whale.

It was a juvenile, about ten years old – João said later – and its pronounced melon meant it was a male; I had learned that the older the animals, the paler they became. But he was still bigger than our boat as he lay there, his greyness shining in the sun.

This time, as the whale came into view underwater, the fear in me subsided as I took in his unbelievable beauty. Forcing my body down, I felt oddly calm. I relaxed; my heart rate began to slow, and I tried to open my eyes wider, to optimize what I could see. Looking into the water, through the sun's rays that played on it from above, I concentrated, committing to memory, even as I saw them, the elements of the whale.

The colour and texture of his skin, shading from smoothness into wrinkled flanks. The rippling muscles, the slatted flukes like an aeroplane's tailfins. His tightly clamped jaw merely made him more placid, playful, even. He did not seem in any hurry to leave. He hung there. And then he turned towards me.

I knew now that the whales had the measure of me; that they knew what I was, even if I could not comprehend them; that I was an object in a four-dimensional map, appraised in six senses. Every nuance of their movement took account of mine. Where I struggled to maintain my balance, to remain part of the encounter, they entirely controlled its choreography.

The young whale moved alongside. Noiselessly, for minutes that seemed like hours, we swam together, eye to eye, fin to fin, fluke to fluke. His movements mirrored my own as we moved in parallel. Black neoprene and grey blubber. Scrawny human and muscled whale. I wasn't afraid any more.

Back in the boat, I watched as the whale turned in a circle. Raising his head one last time, he dipped down, then lifted his flukes, and was gone.

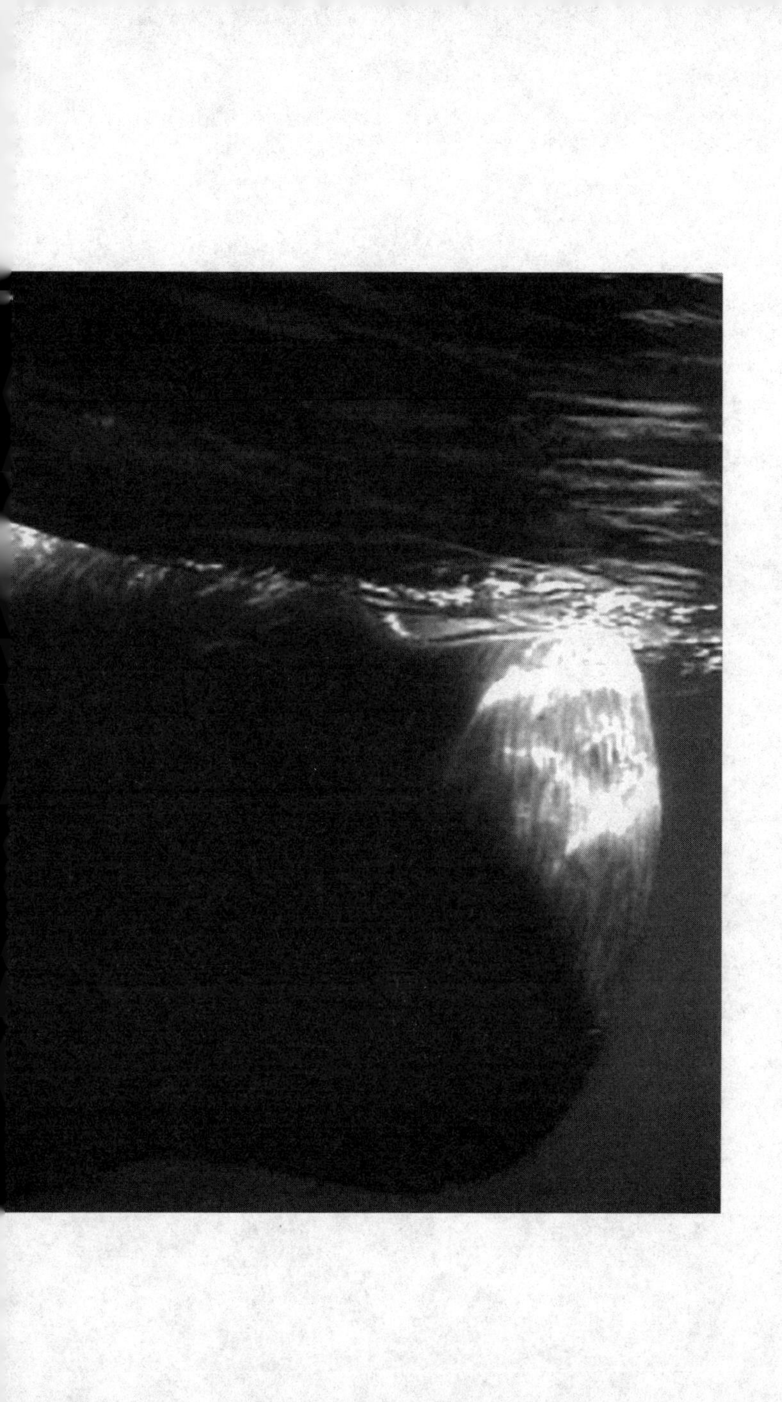

Bibliography

All publications London, unless otherwise stated. For a comprehensive list of sources to the text and further notes, please go to www.harpercollins.co.uk/leviathan

Diane Ackerman, *The Moon by Whale Light*, Orion Publishing, 1993

Peter Adamson, *The Great Whale to Snare: The Whaling Trade of Hull*, Kingston-upon-Hull Museums, Yorkshire (not dated)

Newton Arvin, *Herman Melville*, William Sloane Associates, NYC, 1950

Newton Arvin, editor, *The Heart of Hawthorne's Journals*, Houghton Mifflin, Boston & NYC, 1929

W. H. Auden, *Collected Poems*, Faber, 1976

Thomas Beale, *The Natural History of the Sperm Whale*, J. Van Voorst, 1835

Thomas Beale, *The Natural History of the Sperm Whale*, J. Van Voorst, 1839

Henry Beston, *The Outermost House* Owl Books/Henry Holt, NYC, 1992

A. A. Berzin, *The Sperm Whale*, Jerusalem, 1972

Ray Bradbury, *Green Shadows, White Whale*, HarperCollins, 1992

John Braginton-Smith and Duncan Oliver, *Cape Cod Shore Whaling: America's First Whalemen*, Yarmouth, Mass, 2004

Philip Brannon, *The Picture of Southampton*, (1850), Lawrence Oxley, Alresford, 1973

Frank T. Bullen, *Creatures of the Sea*, Religious Tract Society, 1908

Frank T. Bullen, *The Cruise of the Cachalot*, Smith, Elder, 1910

B. R. Burg, editor, *An American Seafarer in the Age of Sail: The Intimate Diaries of Philip C. Van Buskirk, 1851-1870*, Yale University Press, Connecticut, 1994

Robert Burton, *The Life and Death of Whales*, Andre Deutsch, 1980

Mark Carwardine, *Whales, Dolphins and Porpoises,* Smithsonian Handbooks/Dorling Kindersley, 1995, 2002

Owen Chase, *Shipwreck of the Whaleship Essex*, Lyons Press, NYC, 1999

E. Keble Chatterton, *Whalers and Whaling*, T. Fisher Unwin, 1926

Phil Clapham, *Whales,* WorldLife Library, Scotland, 1997

Nelson Cole Haley, *Whale Hunt*, Mystic Seaport Museum, Connecticut, 2002

James Colnett, RH, *A Voyage to the South Atlantic...for the purposes of extending the Spermaceti Whale Fisheries...*, W. Bennet, 1798

Arthur G. Credland, *The Hull Whaling Trade*, Hutton Press, Yorkshire, 1995

William M. Davis, *Nimrod of the Sea*, Harper & Brothers, NYC, 1874

Daniel Defoe, *A Tour through the Whole Island of Great Britain,* Everyman's, 1966

Andrew Delbanco, *Melville,* Alfred A. Knopf, NYC, 2005

M. Douglas, *Breaking the Record*, Thomas Nelson & Sons, 1902

Frederick Drummer, editor, *The New Illustrated Animal Kingdom*, Odihams, 1959

Richard Ellis, *Monsters of the Sea,* Lyons Press, Connecticut, 1994

Richard Ellis, *The Search for the Giant Squid*, Penguin 1998

John Evelyn, *Diary of John Evelyn*, Everyman, 2006,

Greg Gatenby, *Whales: A Celebration*, Little, Brown, 1983

Oliver Goldsmith, *Animated Nature*, Blackie & Son, 1870

Jonathan Gordon, *Sperm Whales,* WorldLife Library, Minnesota, 1998

Charles Gould, *Mythical Monsters,* (1886), Studio Editions, 1992

Seymour Gross, Edward G. Lueders *et al*, *The Hawthorne and Melville Friendship,* McFarland & Co., North Carolina & London, 1991

Sidney Frederic Harmer and Francis Charles Fraser, *Report on Cetacea stranded on British Coasts,* Longmans/British Museum, 1918

Nathaniel Hawthorne, *The House of the Seven Gables,* Penguin, NYC, 1986

Nathaniel Hawthorne, *Mosses from an Old Manse,* Modern Library Classics, NYC, 2003
Nathaniel Hawthorne, *The Scarlet Letter,* Dover, NYC, 1994
Nathaniel Hawthorne, *Twenty Days with Julian,* New York Review Books, 2003
Mary Heaton Vorse, *Time and the Town: A Provincetown Chronicle,* Rutgers University Press, New Jersey, 1991
Wilson Heflin, *Herman Melville's Whaling Years,* edited by Mary K. Bercaw Edwards and Thomas Farel Heffernan, Vanberbilt University Press, Tennessee, 2004
Bernard Heuvelmans, *In the Wake of Sea Serpents,* Rupert Hart-Davis, 1968
Thomas Hobbes, *Leviathan,* Cambridge University Press, 1991
Miroslav Holub, *Poems: Before and After,* Bloodaxe, 1990
Gordon Jackson, *The British Whaling Trade,* A & C Black, 1978
C. Ian Jackson, editor, *The Arctic Whaling Journals of William Scoresby The Younger,* Hakluyt Society, 2003
C. L. R. James, *Mariners, Renegrades & Castaways: The Story of Herman Melville and the World We Live In,* University Press of New England, Hanover and London, 1978
Henry James, *Hawthorne,* Trent Editions, Nottingham, 1999
Ian Kelly, *Beau Brummell,* Hodder & Stoughton, 2005
D. H. Lawrence, *Studies in Classic American Literature,* Thomas Seltzer, NYC, 1923
John F. Leavitt, *The Charles W. Morgan,* Mystic Seaport Museum, Connecticut, 1998
Jay Leyda, *The Melville Log,* Gordian Press, NYC, 1969
John C. Lilly, *Communication between Man and Dolphin,* Crown, NYC, 1978
Barry Lopez, *Arctic Dreams,* Vintage, NYC, 2001
Andrew Lycett, *Conan Doyle,* Weidenfeld & Nicolson, 2007
Philip McFarland, *Hawthorne in Concord,* Grove Press, NYC, 2004
Leonard Harrison Matthew *et al, The Whale,* Crescent Books, NYC, 1974

James G. Mead and Joy P. Gould, *Whales and Dolphins in Question*, Smithsonian Institution Press, Washington and London, 2002
Herman Melville, *The Whale*, Richard Bentley, 1851
Herman Melville, *Moby-Dick; or, The Whale*, Harper & Brothers, NYC, 1851
Herman Melville, *Moby-Dick*, introduction by Viola Meynell, Oxford University Press, (1920) 1963
Herman Melville, *Moby-Dick; or, The Whale*, Harold Beaver, editor, Penguin, 1972
Herman Melville, *Moby-Dick*, illustrated by Barry Moser, Arion Press/ University of California Press, Los Angeles and London, 1979
Herman Melville, *Pierre, or, The Ambiguities*, Penguin, 1996
Herman Meville, *Redburn: His First Voyage*, Penguin, 1986
Herman Melville, *Typee: A Peep at Polynesian Life*, Penguin, 1996
Herman Melville, *White-Jacket, or, The World in a Man-of-War*, Northwestern University Press, Illinois, 2000
Charles Nordhoff, *Whaling and Fishing*, Dodd, Mead & Company, NYC, 1895 (first published 1856)
Charles Olson, *Call Me Ishmael*, Cape Editions, 1967
J. P. O'Neill, *The Great New England Sea Serpent*, Down East Books, Maine, 1999
George Orwell, *Coming up for Air*, Penguin, 1962
Sonia Orwell and Ian Angus, editors, *The Collected Essays…of George Orwell*, Secker and Warburg, 1968
Vassili Papastavrou, *Eyewitness Whale*, Dorling Kindersley, 2004
Hershel Parker *et al*, *Aspects of Melville*, Berkshire County Historical Society, Pittsfield, Mass, 2001
Hershel Parker, *Herman Melville: A Biography*, Vols I & II, *1851-1891*, Johns Hopkins University Press, Baltimore, 1996 & 2002
The Paris Review Interviews, Vol I, Canongate, 2007
Nathaniel Philbrick, *In the Heart of the Sea*, HarperCollins, 2000
Nathaniel Philbrick, *Mayflower*, Viking Penguin, 2006
Edgar Allan Poe, *The Narrative of Arthur Gordon Pym of Nantucket*, Penguin, 2006

Nicholas Redman, *Whales' Bones of the British Isles,* Redman Publishing, 2004
Randall R. Reeves *et al, Guide to Marine Mammals of the World,* National Audubon Society, Alfred A. Knopf, NYC, 2002
J. Ross Browne, *Etchings of a Whaling Cruise,* (1846), Harvard University Press, Massachusetts, 1968
David Rothenberg, *Thousand Mile Song: Whale Music in a Sea of Sound,* Basic Books, NYC, 2008
Viola Sachs, *The Game of Creation,* Editions de la Maison des sciences de l'homme, Paris, 1982
Victor B. Scheffer, *The Year of The Whale,* Scribner's, NYC, 1969
Sheldrick, M.C., *Stranded whale records, 1967-1986,* Natural History Museum, 1989
Elizabeth A. Schultz, *Unpainted to the Last:* Moby-Dick *and Twentieth-Century American Art,* University Press of Kansas, 1995
R. E. Scoresby-Jackson, *The life of William Scoresby,* 1861
William Scoresby, *An Account of the Arctic Regions,* Constable, Edinburgh, 1820
William Scoresby, *My Father,* 1851
Odell Shepard, *Lore of the Unicorn,* George Allen & Unwin, 1930
Hadoram Shirihai and Brett Jarrett, *Whales, Dolphins and Seals,* A & C Black, 2006
Tom and Cordelia Stamp, *William Scoresby,* Caedom, Yorkshire, 1976
Bram Stoker, *Dracula,* Airmont Publishing, NYC & Toronto, 1965
Thomas Sturge Moore, *Albert Dürer,* Biblio Bazaar, 2007
Algernon Swinburne, *Lesbia Brandon,* Falcon Press, 1952
Henry D. Thoreau, *Cape Cod,* Penguin, NYC, 1987
Henry D. Thoreau, *Walden,* Princeton University Press, New Jersey, 1989
Serge Viallelle, *Dolphins and Whales from the Azores,* Espaço Talassa, Azores, 2002
Howard P. Vincent, *The Trying-Out of Moby-Dick,* Southern Illinois University Press, 1949
Robert K. Wallace, *Douglass and Melville,* Spinner Publications, New Bedford, 2003

Hal Whitehead, *Sperm Whales: Social Evolution in the Ocean*, University of Chicago Press, 2003

Maurizio Würtz and Nadia Repetto, *Dolphins and Whales*, White Star, Vercelli, 2003

NEWSPAPERS, PERIODICALS AND WEBSITES

Associated Press

BBC website

Canadian Journal of Zoology

Daily Mail

Daily Telegraph

The Guardian

The Independent

Historic Nantucket

Illustrated London News

Journal of the House of Commons, british-history.ac.uk

Laelaps, Brian Switek, Rutgers University

'Lost Museum' City University of New York'

Magazine for Natural History, 1835

Natural History

The New York Times online

NRDC Action Fund

Oxford Dictionary of National Biography, online edition

The Pharmaceutical Journal

'Ploughboy', Tom Tyler, Denver University website

Post-Medieval Archaeology

Science News Online

The Scottish Naturalist

Southern Evening Echo, Southampton

Standard-Times, New Bedford

The Times online archive

Times Literary Supplement

Turner Studies, Tate Gallery

FURTHER INFORMATION

'The Hunt for Moby Dick', an Arena film: www.thehuntformobydick.com

UK Whale and Dolphin Stranding Scheme: www.nhm.ac.uk/zoology/stranding

Whale and Dolphin Conservation Society: www.wdcs.org

Provincetown Center for Coastal Studies: www.coastalstudies.org

New Bedford Whaling Museum: www.whalingmuseum.org

International Whaling Commission: www.iwcoffice.org

Picture Credits

8, Katherine Moore; *10,* Jonny Hannah; *40, 178,* Berkshire Atheneum; *47, 161,* Rockwell Kent/R.R. Donnelly & Sons/The Plattsburgh College Foundation; *58–9, 147, 232, 233, 284, 285, 287,* Arthur Credland/Hull Maritime Museum; *64,* Max Goonetillake; *83, 87 a & b, 95, 246, 312,* Natural History Museum, London; *217,* Michael Long; *100, 107, 110, 206, 248,* New Bedford Whaling Museum; *114,* Nantucket Historical Association; *175,* Library of Congress, Washington; *188,* Bodleian Library, Oxford; *196,* Dan Towler; *241, 242–3,* David Connell/ Burton Constable Foundation; *251,* Metropolitan Museum of Art, New York; *271,* Mark Wallinger/Anthony Reynolds; *387,* Antonio Domingos Avila; *394,* Martin Rosenbaum.

All other images from the author's collection.

Acknowledgements

On my third or fourth visit to Provincetown, John Waters accused me of spending more time with whales than with humans; it was perhaps as a form of therapy that he suggested I should write this book. But the roots of my fascination with whales lie with my sister Katherine and her childhood enthusiasm, one which our sister, Christina, shared. Their own children, Oliver, Harriet, Jacob and Lydia, continue that interest – particularly my youngest nephews, Max and Cyrus, both of whom have taught me about whales, despite neither having reached ten years of age. I would like to thank my older brothers, too, Lawrence and Stephen and their families, for their support. As ever, my friend Mark Ashurst has been the ultimate adjudicator of what I do; without him, my book would have been beached long ago.

Leviathan owes its deepest debt to Adam Low and Martin Rosenbaum, director and producer/cameraman respectively of the BBC *Arena* film, *The Hunt for Moby-Dick.* From freezing New England shoots, warmed by nips from Martin's bottle of whisky, to Adam's fearless attempts to direct me and the whales on the high seas despite his propensity to sea-sickness, our adventures together have shaped this book. Adam also read the manuscript and made vital comments. Our guiding light has been *Arena*'s series editor, Anthony Wall; we owe much to his faith and inspiration.

Back at home, Michael Bracewell, Linder Sterling, Neil Tennant, Clare Goddard and Hugo Vickers provided creative encouragement and helpful observations. Liz Jobey published an extract from the book in *Granta* 99; and Keiren Phelan and the Arts Council made possible a late trip to Provincetown. There, Dennis Minsky has been my guide to the world of whale-watching, to the animals' behaviour, and their beauty.

At Fourth Estate, my editor, Mitzi Angel, was true to her name; Nicholas Pearson and Mark Richards provided essential moral and practical support. I would also like to thank Robin Harvie in the publicity department, and Terence Caven, Rachel Smyth and Leo Nickolls for making the book look beautiful. My ever-stalwart agent, Gillon Aitken, steered *Leviathan* to its destination.

Many other people – scientists, curators, writers, historians, librarians, naturalists and artists – have made the journey constantly exciting. Like the whales, they have led me around the world, and are listed here in geographical order.

Provincetown: Charles 'Stormy' Mayo, Jooke Robbins, Scott Landry, Amy Costa, Marc Costa, David Osterberg, Joanne Jarzobski, Nathalie Jacquet, Meribeth Ratzel, Theresa Barbo, Chip Lund, Mary Moore, Ruth Leeney, Beth Swineford, Adam Leiterman, Karen Rankin-Baransky, Karen Stamieskin, Sarah Adams-Fortune, Tanya Gabettie and all the scientists and naturalists and staff at the Center for Coastal Studies; Joe Basine, Mark Delumba and Eric Joranson on the *Portuguese Princess*; the naturalists, captains and crew of the Dolphin Fleet, including Todd Motta, Carol Carlsen, Irene Bragg and John Conlon; the late Frank Schaefer and Mary Martin Schaefer of the legendary White Horse Inn; Mary Oliver for the loan of her field-glasses and dawn discussions on the beach; the late Molly Malone Cook; Pat de Groot for her art and her elegant hospitality; Dan Towler for the postcards; the gentlemen of the Beachcombers Club; Tim Woodman, for his inspiring 'Moby-Dick' paintings; Helen Miranda Wilson, Albert Merola, James Balla, Jackson Lambert, Josiah Mayo, Jody Melander, Jo Hay,

Margery Greenspan, Conny Hatch, Sally Brophy, Pauline Fisher, and Debbie Minsky for their friendship.

New Bedford: Stuart M. Franks, Mary K. Bercaw Edwards and Arthur Motta for their observations on Melville, whaling, and the *Charles W. Morgan*; Kathy Reed of Edgewater for her accomodation. **Nantucket:** Nantucket Historical Association, Whaling Museum and Research Library. **Western Massachusetts**: Louise McCue and Bobbie-Anne Fachini at Arrowhead, Kathleen Reilly and Ann-Marie Harris at the Berkshire Atheneum. **Connecticut:** Megan Wilson and Duncan Hannah for their company climbing Monument Mountain; William Peterson and the staff of Mystic Seaport Museum. **New York:** Jack Puttnam for his tour around Melville's Manhattan; Thomas Farel Heffernan, Dan and Lucia Woods Lindley; and Richard Melville Hall for tea and directions to Herman Melville Place. **Maine:** Alex Carleton, for his aesthetic inspiration, and all at Rogues Gallery, especially Daniel Pepice. **Newfoundland:** Hal Whitehead, for sharing a little of his work on sperm whales.

Andalucia: Gabriel Orozco for a tour of his tattooed whale; José María Galán and the Museo Marino, Matalascañas. **Azores:** Serge Viallelle for introducing me to the sperm whales; Alexandra Viallelle, João Quadresma and Marco Avila and all at Espaço Talassa, Lajes do Pico; Macolm Clarke for his commentary on *Physeter* and their prey; Dorothy Clarke; Antonio Domingos Avila; Museu dos Baleerios, Pico; Museo da Industria Baleeira, São Roque. A special thanks to the Azorean government and the regional secretary for the environment for licencing my close encounter with *Physeter macrocephalus.*

London: Richard Sabin, for his patient answers to my many questions; Liz Evans-Jones at the Strandings Project; Stephen Roberts, Becci Cousins, Katie Andersen, Polly Tucker, Helen Sturge and the staff of the Natural History Museum and its archives; Lisa Le Feuvre and Helen

Whiteoak at the National Maritime Museum; James Rawlinson, Richard Mortimer, Diane Gibbs and Christine Reynolds at Westminster Abbey Museum and Library; the staff of the Guildhall Library and the British Library; Gilbert and George, Jeremy Millar, Tim Marlow, Honey Luard, Anthony Reynolds, Michael Prodger, Giles Foden, Boyd Tonkin, Simon Callow, Reed Wilson, Madeleine Groves, Michael Holden, Julia Harrison, Nicholas Redman, Peter David, Steve Deput, Sam Goonetillake, Namvula Rennie, and Emma Matthews.

Yorkshire: John Chichester-Constable, David Connell and Gary Dewson at Burton Constable Hall; Dr Michael Boyd; and particular thanks to Arthur Credland and the Hull Maritime Museum; Whitby Museum. **Oxford**: Paul Bonaventura, Ruskin School of Art; Malgosia Nowak-Kemp, Oxford University Museum of Natural History, Clive Hurst and the staff of the Rare Books and Printed Ephemera, the Bodleian Library. **Devon**: Nigel Larcombe-Williams, Jake Luffman. **Hampshire**: Peter Leslie, Jude James, Colin Speedie, Clare Moore. **Southampton:** Sophia Scott and Alison Kentuck at HM Maritime and Coastguard Agency; Southampton City Library; Tina Jones; Andy and Rob, Sholing Cycle Centre; Fr Bill Wilson, Katherine Anteney, Jonny Hannah, and Pamela and Ron Ashurst. I'd also like to thank to Krishna Stott, Jon Wynne-Tyson, D.J. Taylor, Jonathan Gordon and all those who have contributed to a story which, I hope, will have a happier ending than its beginning.

Philip Hoare,
Southampton, July 2008

Index

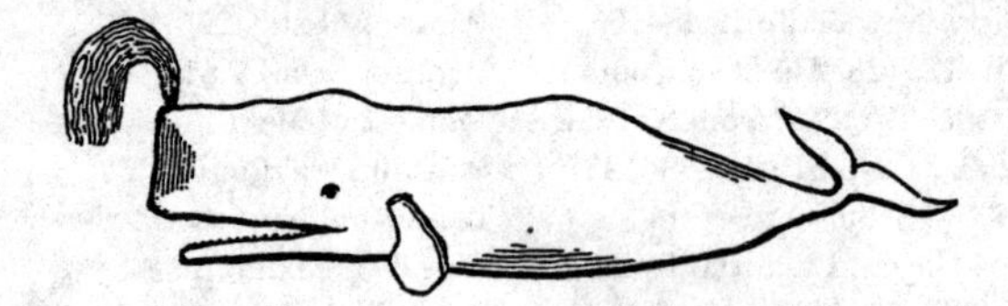